ENVIRONMENTAL HAZARDS OF METALS

Toxicity of Powdered Metals and Metal Compounds

STUDIES IN SOVIET SCIENCE

PHYSICAL SCIENCES

1973

Densification of Metal Powders during Sintering
V. A. Ivensen

The Transuranium Elements
V. I. Goldanskii and S. M. Polikanov

Gas-Chromatographic Analysis of Trace Impurities
V. G. Berezkin and V. S. Tatarinskii

A Configurational Model of Matter
G. V. Samsonov, I. F. Pryadko, and L. F. Pryadko

Complex Thermodynamic Systems
V. V. Sychev

Crystallization Processes under Hydrothermal Conditions
A. N. Lobachev

Migration of Macroscopic Inclusions in Solids
Ya. E. Geguzin and M. A. Krivoglaz

1974

Theory of Plasma Instabilities
Volume 1: Instabilities of a Homogeneous Plasma
A. B. Mikhailovskii

Theory of Plasma Instabilities
Volume 2: Instabilities of an Inhomogeneous Plasma
A. B. Mikhailovskii

Nonequilibrium Statistical Thermodynamics
D. N. Zubarev

Refractory Carbides
G. V. Samsonov

Waves and Satellites in the Near-Earth Plasma
Ya. L. Al'pert

1975

Environmental Hazards of Metals: Toxicity of Powdered Metals and Metal Compounds
I. T. Brakhnova

STUDIES IN SOVIET SCIENCE

ENVIRONMENTAL HAZARDS OF METALS

Toxicity of Powdered Metals and Metal Compounds

I. T. Brakhnova
Materials Science Institute
Academy of Sciences of the UkrSSR
Kiev, USSR

Translated from Russian by

J. H. Slep

CONSULTANTS BUREAU • NEW YORK AND LONDON

Library of Congress Cataloging in Publication Data

Brakhnova, Irina Tikhonovna.
Environmental hazards of metals.

(Studies in Soviet science)
Translation of Toksichnost' poroshkov metallov i ikh soedinenii.
Bibliography: p.
1. Metal powders—Toxicology. 2. Powder metallurgy—Hygienic aspects. 3. Metal-workers—Diseases and hygiene. I. Title. II. Series.
RA1231.M52B6813 615.9'25'3 75-11921
ISBN 0-306-10897-6

Irina Tikhonova Brakhnova was born in 1925. In 1952 she was graduated from the Health and Hygiene Department of the I. M. Sechenov First Moscow Medical Institute. She is presently the director of the laboratory of industrial aerosols at Kiev Institute of Industrial Hygiene and Occupational Diseases.

The original Russian text, published by Nauka Dumka in Kiev in 1971, has been corrected by the author for the present edition. This translation is published under an agreement with the Copyright Agency of the USSR (VAAP).

ТОКСИЧНОСТЬ ПОРОШКОВ МЕТАЛЛОВ И ИХ СОЕДИНЕНИЙ
И. Т. Брахнова
TOKSICHNOST' POROSHKOV METALLOV I IKH SOEDINENII
I. T. Brakhnova

A Division of Plenum Publishing Corporation
227 West 17th Street, New York, N.Y. 10011

United Kingdom edition published by Consultants Bureau, London
A Division of Plenum Publishing Company, Ltd.
Davis House (4th Floor), 8 Scrubs Lane, Harlesden, London, NW10 6SE, England

Printed in the United States of America

Foreword

Metal powders are solid materials which are distinguished from large pieces of metals by their small size and by the immensel high ratio of surface to volume. Their large specific surface plays an important role wherever surface reactions take place. Densification and shrinkage of a powder mass depends to a large extent on the specific surface area.

The large specific surface of powders also contributes to the development of several potentially hazardous properties such as pyrophoricity (flammability), explosivity, and toxicity. Much information has been published concerning the pyrophoricity and explosivity of metal powders. However, with respect to the toxicity of metal powders, Miss Brakhnova's book is unique: it is the most comprehensive book on this subject.

There are dangerous *properties;* there are no dangerous *materials.* Some materials can be hazardous to the uninformed, can have properties that are dangerous if misapplied. They present no danger at all as long as we are aware of these properties and know how to handle them. The key to their proper use is to be aware of such characteristics. It is to the great credit of Miss Brakhnova that she has called some potentially hazardous properties of metal powders to our attention.

The author treats the subject from a fundamental as well as from a practical aspect; she discusses not only metals but also a great variety of metal compounds. The first chapter presents a concise review of the electronic structure of atoms and the crystal-chemical properties of metallic materials which will be of interest to a wide variety of readers. The introduction to the general problems of industrial hygiene that follows will be most revealing and useful to the majority of powder metallurgists.

The toxic effects of powders of common metals and also of refractory compounds such as oxides, carbides, borides, silicides, nitrides, and others form the content of the subsequent chapters. We are well aware that the Academy of Sciences of the Ukrainian SSR in Kiev is one of the leaders in research and development of refractory compounds and we refer to the numerous books of Academician G. V. Samsonov, with whom Miss Brakhnova worked in close cooperation.

An extensive chapter (VIII) deals with the theoretical problems of toxicology in general and especially with the toxicity of metallic materials. Warnings of this kind are most valuable. Even more valuable is the fact that the author presents information about measures to be taken for the prevention of occupational diseases among workers exposed to certain metallic powders.

The information given in this book is based on the author's own experience and on more than 500 references, of which approximately 60% are from the Russian and approximately 40% from the Western literature.

The book is of great value not only to powder metallurgists, ceramists, and other powder technologists, but also to chemists and industrial hygienists and is highly recommended for extensive study.

New York
April 1975

Henry H. Hausner
Adjunct Professor
Polytechnic Institute of
New York

Preface

Metal powders and metallic compounds, particularly oxides, are used widely as alloying additives in ferrous and nonferrous metallurgy, are the principal raw material in the electrode industry, and are employed extensively as pigments in the manufacture of porcelain, majolica, and faience ware and in the glass industry. During various production operations they can be the source of formation of condensation and disintegration aerosols and have an adverse effect on workers. A broad assortment of finely divided metallic materials is used in powder metallurgy plants in the manufacture of sintered articles.

High wear resistance of sintered articles, heat resistance, refractoriness, hardness, corrosion resistance, and other valuable properties in powder metallurgy are attained by the use not only of various metal but also of new powder materials. A special place among them is occupied by refractory compounds (borides, carbides, silicides, nitrides) and chalcogenides (sulfides, selenides, tellurides), which are used primarily as semiconducting materials, and also metal carbonyls, which are used for the manufacture of high-purity metal powders and as catalysts of chemical processes.

Problems of the toxic effect of aerosols of individual metals have been elucidated rather fully in Soviet and foreign literature, and worthy of particular attention is this respect are the investigations carried out under the supervision of Z. I. Izraél'son at the department of hygiene of the I. M. Sechenov First Moscow Medical Institute (O. Ya. Mogilevskaya, I. V. Roshchin, N. V. Mezentseva, Z. S. Kaplun, R. V. Borisenkova, et al.). The biological activity of powdered metallic mixtures and new compounds that are acquiring ever greater importance has been studied less.

To prevent diseases it is important not only to determine the degree of danger associated with the action of various materials on the organism, but also to lay down guidelines for a preliminary evaluation of new chemical substances being introduced into industry, whose proper toxicological investigation cannot be carried out in a short time. Data yielded by a comparison of the toxicity of various metallic aerosols with their electronic structure and crystal-chemical properties can be used for this purpose.

One of the most prominent theoretical physicists of our day, R. Feynman, in his book "Character of Physical Laws," indicates that "the most fruitful thought most strongly stimulating progress in biology is, apparently, the conjecture that everything that animals do atoms do, too, and that in living nature everything is the result of some kind of physical and chemical processes and there is nothing over and above this." Many authors have investigated the degree of activity of numerous biologically active substances as a function of their electronic structure and have made the pertinent quantum-mechanical calculations (C. Reed, 1960; A. Szent-Györgi, 1964; B. and A. Pullman, 1965, et al.).

This book presents the results of our investigations at the Kiev Institute of Labor Hygiene and Occupational Diseases and the literature data on the conditions of formation of metallic aerosols in powder metallurgical plants and in allied branches of industry, the state of health and morbidity among workers exposed to metallic aerosols, and also experimental data on the fibrogenic and general toxic effects of the principal powdered materials used in the powder metallurgy industry.

An attempt is made to relate the activity of such substances to their electronic and crystal structures, and this permits us to outline the basic laws governing the toxic action of substances and compounds. The pathogenic activity of simple substances, oxides, refractory compounds, and chalcogenides, and also certain complex compounds, particularly carbonyls, is examined from this viewpoint. For these reasons, it was necessary to present existing data on the electronic and crystal structure of metals and their compounds before dealing with the subject matter stated in the title. These data will allow the reader to understand more thoroughly the causes of formation of metallic aerosols in an industrial setting.

The author wishes to thank G. V. Samsonov, L. N. Bazhenova, Z. I. Izraél'son, L. I. Medved', and E. I. Chaika for consultation and advice during the investigations and work on the monograph, and also N. V. Lazarev and V. A. Obolonchik for reviewing the manuscript.

Contents

Chapter I. Electronic Structure and Crystal-Chemical Properties of Metals and Their Compounds 1

Electronic Structure of Atoms 1
Characteristics of the Structure of Metal Atoms 7
Crystal-Chemical Characteristics of Metals and Their Compounds 11
Complex Compounds of Metals and the Formation of a Coordination Bond 32

Chapter II. Conditions of Formation of Metallic Aerosols in the Air of Industrial Rooms in Powder Metallurgy and Allied Plants . . 39

Industrial Hygiene in the Production of Refractory Compounds 40
Hygienic Characteristics of Working Conditions in the Production of Iron Powders 47
Conditions of Formation of Metallic Aerosols during the Manufacture of Sintered Articles 53

Chapter III. Morbidity and State of Health of Workers Exposed to Metallic Dust 71

Morbidity with Temporary Disability among Workers Exposed to Metallic Aerosols 71
State of Health of Workers Exposed to Metallic Dust 74

Chapter IV. Toxic Effect of Some Metal Powders and Powder Metallurgy Mixtures . . . 93

Hygienic Evaluation of the Effect on the Organism of Copper Powder and Some Copper-Base Powder Metallurgy Mixtures 93
Toxic Effect of Dust of Iron Oxides and Iron Powders Produced by Reduction 99
Pneumoconiotic Changes and General Toxic Effect of Carbonyl Iron Powders 103
Toxicity Characteristics of Some Powder Metallurgy Composites of Iron with Graphite and Nonferrous Metals 108
Fibrogenic and General Toxic Effect of Some Powdered Ferrite Composites 111

Chapter V. Toxic Effect of Refractory Compounds 115
Hygienic Evaluation of the Effect of Transition Metal Borides and Carbides on the Organism 115
Comparative Hygienic Evaluation of the Toxic Effect of Transition Metal Silicides 130
Effect of Refractory Nitrides on the Organism 143
Toxic Effect of Rare Earth Refractory Compounds 151

Chapter VI. Toxic Effect of Some Chalcogenides 161
Toxic Effect of Some Metallic Sulfides 163
Toxic Effect of Selenides and Tellurides of Transition Metals 165

Chapter VII. Toxic Effect of Transition Metal Carbonyls 179

Chapter VIII. Theoretical Problems of the Toxicology of Metals and Their Compounds 191
Variation of the Toxic Properties of Simple Substances as Function of Their Electronic Structure 194
Regularities of the Toxic Effect of Transition Metal Oxides 204
Dependence of the Toxic Effect of Compounds of Metals with Nonmetals on Their Electronic Structure 210
Toxic Effect of Some Complex Compounds 222

Relation between the Formation of Metallic Aerosols and Their Electronic and Crystal Structures 227

Chapter IX. Measures for Preventing Occupational Diseases among Workers in Powder Metallurgy Plants 231

General Requirements Imposed on Industrial Organization. 233
Production of Sintered Articles from Refractory Compounds 235
Production of Iron Powders by Reduction 237
Production of Carbonyl Iron Powders 239
Some Recommendations on Performing the Main Production Processes Accompanied by Dust Liberation 240
Medical Service for Workers 246

Conclusion 251

Literature Cited 255

CHAPTER I

Electronic Structure and Crystal-Chemical Properties of Metals and Their Compounds

Electronic Structure of Atoms

In 1912 Niels Bohr suggested that electrons revolve around a nucleus only in certain "allowed" orbits. Following the idea of Max Planck concerning quantization of energy, he postulated that the change of energy upon absorption and radiation of light is of a quantum character and he introduced the notion of discrete stationary orbits, thereby laying the foundation for the modern physics of electron shells.

Modern quantum mechanics—the mechanics of "wave particles"—is based on the idea that an electron has a wave nature. The allowed orbits of the Bohr quantum theory correspond to allowed "states" in the wave theory. A system of quantum numbers n, l, m_l, s corresponds to discrete values of the energy of an electron in stationary states or in stable orbits.

The principal quantum number n determines the electron energy and can take on only positive integral values, n = 1, 2, 3, 4, 5, . . . ; the greater the value of n, the higher the electron energy.

The orbital quantum number l determines the value of the orbital angular momentum of the electron, i.e., it characterizes the form of the shell. It can take on values of 0, 1, 2, 3, . . . , up to a maximum value given by $l = n - 1$.

The atom and electron possess magnetic properties which have an effect on their state. The orbital motion of the electron gives rise to a magnetic moment, and the third quantum number m_l, the so-called orbital magnetic moment or magnetic quantum number, is introduced to characterize the effect.

The magnetic quantum number m_l determines the magnitude of the projection of the orbital momentum of the electron along a given direction and also the projection of the orbital magnetic moment of the electron in the direction of a weak external magnetic field. The magnetic quantum number can have the values $m_l = 0, \pm 1, \pm 2, \ldots, \pm l$, i.e., it can take on $2l + 1$ values.

It is known that electrons not only revolve in orbits but also rotate about their own axes. The quantum number s determines the value of the projection of the rotational moment of the electron, or its spin, on a given direction and is equal to $+1/2$ or $-1/2$. Some electrons of the orbit rotate in a mutually opposed paired state. The number of unpaired, or valence, electrons of an atom can be determined by representing the configuration of the atom by means of energy cells (see diagram; the meaning of the designations 1s, $2p^3$, etc., and the rationale for this diagram are given below):

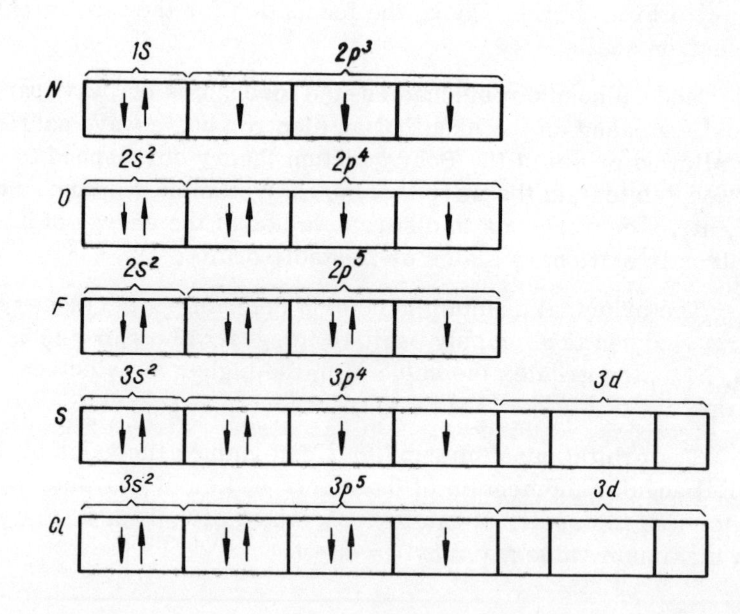

The nitrogen atom has three unpaired electrons, the oxygen and sulfur atoms two each, and the fluorine and chlorine atoms one each.

The radii of the electron orbits are related to one another as $1^2:2^2:3^2:4^2:n^2$. The principal quantum number gives the number of the "allowed" (Bohr) orbit or electron shell, e.g., n = 1 corresponds to the first, or innermost, shell. The electrons which surround the nucleus are thus located in several electron shells, the electrons furthest from the nucleus being most weakly bound to it. The electrons of the last or penultimate shell are therefore the most mobile, and only they can participate in the formation of a chemical bond between atoms. From this it follows that the chemical properties of elements are determined mainly by the structure of the outer electron shells of their atoms.

There is a limit to the number of electrons that can "fit" into each shell (see below); when one shell is filled, additional electrons must go into the next outer shell. Electrons in the case of a free unexcited state of the atom must first fill the lowest energy levels (Fig. 1). Thus the maximum principal quantum number corresponds to the number of the outer electron shell of the atom of a given element and is equal to the number of the period to which it belongs in the periodic table of the elements.

The principal quantum number n is denoted by numbers (n = 1, 2, 3, etc.) or by letters K (n = 1), L (n = 2), M (n = 3), etc. The azimuthal (or orbital) quantum number l is also denoted by small Latin letters s (sharp), p (principal), d (diffuse), and f (fundamental). In designating the electron composition of atoms, the principal quantum number n, which gives the number of the shell occupied by the electron, is written before the letter indicating the value of the quantum number l: s (for $l = 0$), p (for $l = 1$), d (for $l = 2$), or f (for $l = 3$). The number of electrons in a given energy level is indicated by an exponent. For example, $2p^3$ designates a shell with n = 2, $l = 1$, and occupied by three electrons.

The Swiss physicist Pauli put forward the hypothesis that no two electrons in an atom can have the same values for all four of the quantum numbers. The maximum number of electrons in a given shell is then determined by the equation

$$N = 2\sum_{l=1}^{n-1}(2l+1) = 2n^2,$$

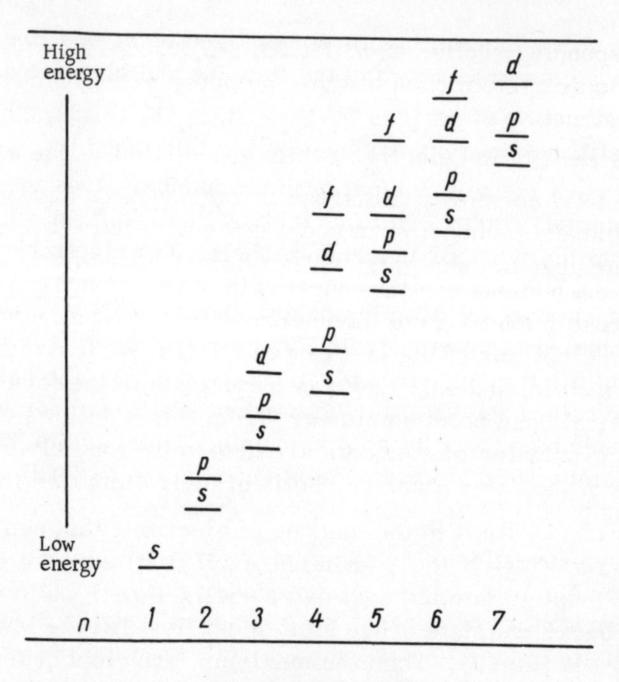

Fig. 1. Energy levels.

which can be seen as follows: The magnetic quantum number corresponds to $2l + 1$ states of the electrons in the atom, and the fourth quantum number s can have two values, $\pm\frac{1}{2}$. Thus the number of electrons in an orbit corresponding to the principal quantum number n and orbital quantum number l is equal to $2(2l + 1)$. Summation over the allowed values of l for a given n yields the above equation.

For example, we have $l = 0$ when $n = 1$ (K shell). The maximum number of electrons corresponding to the first shell is thus equal to $2(2l + 1) = 2(2 \times 0 + 1) = 2$, i.e., on the first shell and hence in the first period there are only two electrons ($1s^2$). In the second, L, shell, we can have $l = 1$ and $2(2 \times 1 + 1) = 6$. Since both $l = 0$ and $l = 1$ are possible for the L shell, there can be two s electrons in the first subshell ($l = 0$) and six p electrons in the second subshell ($l = 1$). If the L shell is filled, we can write its electronic configuration as $2s^2 2p^6$.

Substituting the principal quantum number n = 1, 2, 3, 4, 5, 6, into the above equation we obtain the number of electrons in the corresponding shells: 2, 8, 18, 32, 50, 72, . . . , $2n^2$. Actually, in the periodic system each of these numbers except 2 is repeated in the construction of periods I-VII: 2, 8, 8, 18, 18, 32, 32, ... There is still no rigorous explanation for this fact.

The total number of electrons in each shell is distributed among sublevels (subshells) in accordance with the above rules for the four quantum numbers: s^2, s^2p^6, $s^2p^6d^{10}$, $s^2p^6d^{10}f^{14}$, ...

As a study of the atomic spectra showed, filling of the subshells occurs in accordance with Hund's rule. At first one half of the shell is filled by unpaired electrons, and during the filling of the second half of the shell the subsequent electrons form pairs having coupled antiparallel spin moments. This results in not only the completely filled electron shells but also the half-filled configurations p^3, d^5, f^7 displaying definite stability.

Many properties of substances are governed by the structure of the outer electron shells. Atoms having two or eight electrons in the outer shells are distinguished by great stability. This is characteristic of inert gases, which for all practical purposes do not interact with other elements. Other elements in their compounds try to acquire the structure of the nearest inert gas. The number of electrons in the outer shell increases from left to right in the short periods of Mendeleev's system. Alkali metals have one s electron each; alkaline earth metals have two s electrons each, etc; halogens have seven s^2p^5 electrons.

In the long periods the penultimate d shell rather than the outer shell is filled in the intervals Sc–Zn and Y–Cd. The outer shell of these elements remains unchanged.

The electrons of the penultimate shell are bound more strongly with the nucleus in comparison with the outer shell electrons, and therefore they have less mobility. Thus, such electrons determine the chemical properties of substances to a lesser degree and differ less from one another.

In the sixth period there are, in addition to the unfilled d shell, 14 elements with an unfilled third-from-outside f shell, for which its maximum filling to 14 electrons is characteristic. Since this shell is located still closer to the nucleus and the electrons in it are still less mobile, the differences between the elements here are even smaller.

TABLE 1. Electronic Structure of Isolated Atoms of Metals

Atomic number	Element	K	L		M			N				O				P
		1s	2s	2p	3s	3p	3d	4s	4p	4d	4f	5s	5p	5d	5f	6s
	Group I (s metals)															
3	Li	2	1	—	—	—	—	—	—	—	—	—	—	—	—	—
11	Na	2	2	6	1	—	—	—	—	—	—	—	—	—	—	—
19	K	2	2	6	2	6	—	1	—	—	—	—	—	—	—	—
37	Rb	2	2	6	2	6	10	2	6	—	—	1	—	—	—	—
55	Cs	2	2	6	2	6	10	2	6	10	—	2	6	—	—	1
4	Be	2	2	—	—	—	—	—	—	—	—	—	—	—	—	—
12	Mg	2	2	6	2	—	—	—	—	—	—	—	—	—	—	—
20	Ca	2	2	6	2	6	—	2	—	—	—	—	—	—	—	—
38	Sr	2	2	6	2	6	10	2	6	—	—	2	—	—	—	—
56	Ba	2	2	6	2	6	10	2	6	10	—	2	6	—	—	2
	Group II (d metals)															
21	Sc	2	2	6	2	6	10	2	—	—	—	—	—	—	—	—
39	Y	2	2	6	2	6	10	2	6	1	—	2	—	—	—	—
22	Ti	2	2	6	2	6	2	2	—	—	—	—	—	—	—	—
40	Zr	2	2	6	2	6	10	2	6	2	—	2	—	—	—	—
72	Hf	2	2	6	2	6	10	2	6	10	14	2	6	2	—	2
23	V	2	2	6	2	6	3	2	—	—	—	—	—	—	—	—
41	Nb	2	2	6	2	6	10	2	6	4	—	1	—	—	—	—
73	Ta	2	2	6	2	6	10	2	6	10	14	2	6	3	—	2
24	Cr	2	2	6	2	6	5	1	—	—	—	—	—	—	—	—
42	Mo	2	2	6	2	6	10	2	6	5	—	1	—	—	—	—
74	W	2	2	6	2	6	10	2	6	10	14	2	6	4	—	2
25	Mn	2	2	6	2	6	5	2	—	—	—	—	—	—	—	—
43	Tc	2	2	6	2	6	10	2	6	5	—	2	—	—	—	—
75	Rc	2	2	6	2	6	10	2	6	10	14	2	6	5	—	2
26	Fe	2	2	6	2	6	6	2	—	—	—	—	—	—	—	—
44	Ru	2	2	6	2	6	10	2	6	7	—	1	—	—	—	—
76	Os	2	2	6	2	6	10	2	6	10	14	2	6	6	—	2
27	Co	2	2	6	2	6	7	2	—	—	—	—	—	—	—	—
45	Rh	2	2	6	2	6	10	2	6	8	—	1	—	—	—	—
77	Ir	2	2	6	2	6	10	2	6	10	14	2	6	7	—	2
28	Ni	2	2	6	2	6	8	2	—	—	—	—	—	—	—	—
46	Pd	2	2	6	2	6	10	2	6	10	—	0	—	—	—	—
78	Pt	2	2	6	2	6	10	2	6	10	14	2	6	9	—	1
29	Cu	2	2	6	2	6	10	1	—	—	—	—	—	—	—	—
47	Ag	2	2	6	2	6	10	2	6	10	—	1	—	—	—	—
79	Au	2	2	6	2	6	10	2	6	10	14	2	6	10	—	1
30	Zn	2	2	6	2	6	10	2	—	—	—	—	—	—	—	—
48	Cd	2	2	6	2	6	10	2	6	10	—	2	—	—	—	—
80	Hg	2	2	6	2	6	10	2	6	10	14	2	6	10	—	2
	Group III (f metals)															
57	La	2	2	6	2	6	10	2	6	10	—	2	6	1	—	2
58	Ce	2	2	6	2	6	10	2	6	10	2	2	6	—	—	2
59	Pr	2	2	6	2	6	10	2	6	10	3	2	6	—	—	2
60	Nd	2	2	6	2	6	10	2	6	10	4	2	6	—	—	2
61	Pm	2	2	6	2	6	10	2	6	10	5	2	6	—	—	2
62	Sm	2	2	6	2	6	10	2	6	10	6	2	6	—	—	2
63	Eu	2	2	6	2	6	10	2	6	10	7	2	6	—	—	2
64	Gd	2	2	6	2	6	10	2	6	10	7	2	6	1	—	2
65	Tb	2	2	6	2	6	10	2	6	10	8	2	6	1	—	2
66	Dy	2	2	6	2	6	10	2	6	10	10	2	6	—	—	2
67	Ho	2	2	6	2	6	10	2	6	10	11	2	6	—	—	2
68	Er	2	2	6	2	6	10	2	6	10	12	2	6	—	—	2
69	Tu	2	2	6	2	6	10	2	6	10	13	2	6	—	—	2
70	Yb	2	2	6	2	6	10	2	6	10	14	2	6	—	—	2
71	Lu	2	6	10	2	6	10	2	6	10	14	2	6	1	—	2

The Pauli principle made it possible to determine the number of electrons in the shells, but it did not make it possible to explain the order of filling the electron shells; for example, it was unclear why in cerium ($Z = 58$) it is the 4f and not the 5d subshell that is filled. It was established later that this order of filling of the electron shells is energetically more advantageous.

Characteristics of the Structure of Metal Atoms

Elements having metallic properties can be divided into three main groups: elements whose s shell fills up, elements with an incomplete d shell, and elements whose f shell is unfilled. The electronic structure of the metals of the three groups is presented in Table 1.

The alkali metals of Group I of the periodic system are characterized by the presence of one s electron in the outer shell, which is over the inert gas shell strongly shielding the nucleus. Therefore, it is weakly bound with the nucleus and can be transferred easily upon interaction with other elements. However, alkali metals exhibit substantial differences as well as definite similarities. For instance, lithium and sodium interact with water and halogens much more weakly than do potassium, rubidium, and cesium.

Lithium, like hydrogen, has one s electron in the outer shell, but in lithium it is shielded from the nucleus by the $1s^2$ shell, whereas in hydrogen the s electron occupies the 1s level, i.e., is bound directly with the nucleus. In the other alkali metals – sodium, potassium, rubidium, cesium, and francium – the outer s electron is shielded also by filled s^2p^6 shells.

The metals of Group IIA have two s electrons in the outer s shell ($2s^2$) which also lie over strongly shielding s^2 and s^2p^6 shells.

The two outer electrons of beryllium located at the 2s level have antiparallel spins, but their attraction to the nucleus is greatly weakened as a consequence of the presence of two electrons in the deeper 1s shell. Beryllium is characterized by the possibility of $s \rightarrow p$ transitions with the formation of sp states along with s^2 configurations, which gives rise to its amphoteric properties.

The probability of this transition decreases for Mg, and it becomes practically impossible for Ca, Sr, and Ba in connection

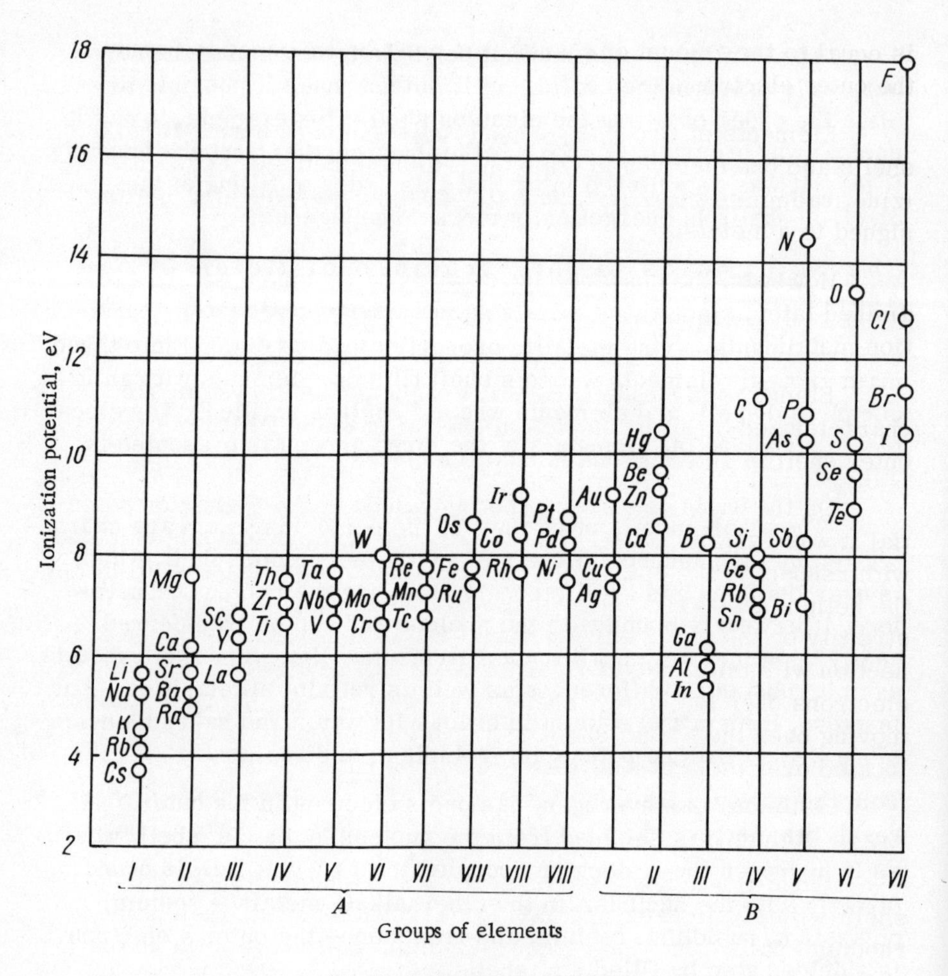

Fig. 2. Ionization potentials.

with the decrease of the energy stability of the sp configuration as the principal quantum number increases. The $2s^2$ electrons of the lower-lying elements, in which they are shielded not only by the s shells but also by the deeper s^2p^6 shells, are bound with the nucleus still more weakly. The outer electrons of Mg are bound more strongly with the nucleus than those of Ca, Sr, Ba, since the weakly shielding $1s^2$ shell has a substantial effect on the strength of this bond.

A measure of the strength of this bond and accordingly of the "metallicity" of an element is the ionization potential, which is equal to the amount of energy that must be expended to remove the outer electrons (Fig. 2).

Elements of Group IB (copper, silver, gold), having filled d shells and one electron in the s shell and elements of Group IIB (zinc, cadmium, mercury), with two s electrons, can also be assigned to s metals.

However, the characteristics of their structure will be examined later, since these metals can be assigned also to d-transition metals.

Elements of Group IIIB–VIIIB of the periodic system are characterized by an incomplete d shell, for which filling of the outer shell to 10 electrons is characteristic.

On the basis of studying the magnitude of the ionization potential it was established [81] that the charge of the nucleus is shielded with respect to the outer valence s and p electrons least of all by the helium shell ($1s^2$), more strongly by the d^{10} shell, and most strongly by the s^2p^6 shell, i.e., the shell of inert gases. In connection with this, maximally bound with the nucleus are the valence electrons over the helium shell, more weakly the valence electrons moving over the $s^2p^6d^{10}$ shell, and still more weakly the electrons located over the s^2p^6 shell, i.e., the farther the electron shell is from the nucleus, the more strongly it is shielded and the more weakly its electrons are bound with the nucleus.

Since 10 elements correspond to maximum filling of the d shell (with 10 electrons), triads (iron–cobalt–nickel, rubidium-rhodium–palladium, and osmium–iridium–platinum) occur in one of the eight groups. The distribution of transition metals between the groups is determined by the total number of d and s electrons in the outer unfilled shells, and only for the cobalt and nickel subgroup is it equal to nine and ten, respectively, thus differing from the number of the group (VIII) of the periodic system.

Weakening of the attraction of the valence electrons by the nucleus is especially noticeable for metals of period VI owing to the so-called lanthanide contraction as a consequence of filling of the 4f shell of these elements.

In the case of the lanthanide contraction an increase of the atomic charge weakly affects the force of attraction of the outer electron shell by the nucleus and thus is not accompanied by a noticeable increase of the atomic radius. This occurs as a consequence of the shielding action of the filled 4d shell.

A special feature of the electronic structure of metals of Group III – the f metals or lanthanides – consists in the successive filling of the f shell, beginning with cerium and ending with lutetium (see Table 1). At present lanthanum is also included among the lanthanides, although spectroscopic data show that filling of the f shell is noted for cerium but not for lanthanum. That it belongs to the lanthanides is confirmed by the similarity between its properties and those of these metals. In connection with this, all lanthanides are placed in a single square of the periodic table.

A characteristic feature of the f elements is the great similarity of their physicochemical and other properties. However, the rare earth elements also have marked differences.

All lanthanides are divided into two groups on the basis of their properties: the cerium group (Ce, Pr, Nd, Pm, Sm, Eu, Gd) and the yttrium group (Tb, Dy, Ho, Er, Tu, Yb, Lu). This division was originally established on the basis of an insignificant difference of properties. Later this division was substantiated by results of physical observations.

The difference between the first and second septets of electrons consists in that in the cerium group the seven electrons have parallel spins and in the yttrium group there is a further filling of the f level to 14 electrons which are arranged in the same quantum cells and have an opposed spin. Paired electrons with antiparallel spins are formed. The direction of natural spin of the first and second electrons is opposite.

In connection with the indicated features of the electronic structure, the rare earth elements are characterized by the presence of a secondary periodicity: In the yttrium group lutetium, having a filled f shell, f^{14}, possesses the strength of the shell of an "inert gas," like gadolinium in the cerium group.

The periodic change of the properties of lanthanides with change of atomic number enabled V. K. Grigorovich [81] to arrange them in appropriate groups of the periodic system, after having separated out three additional subgroups.

A characteristic feature of lanthanides is the ability of their f electrons to undergo f $\rightarrow$ d transitions, which is due to the closeness of the energy states of the f and d levels. Thanks to this, f elements virtually behave as d elements in reactions. However, their physicochemical characteristics are comparatively low, and as a consequence of this the character of change of the physical properties of f elements is similar to that of d elements. In both groups completion of the electron shells occurs: the outer in the case of d-transition elements and the penultimate in the case of the f elements.

The f elements differ from the d elements mainly in the range of variation of properties. Whereas there are substantial differences in the properties of the d elements, the variations in properties of the f elements are small, which is closely related with the change of the atomic radii of the rare earth elements. It was found that the atomic and ionic radii decrease with an increase of the atomic number of f metals.

Differences between d and f elements are manifested also in a tendency towards a change in the degree of oxidation. A large number of degrees of oxidation is characteristic of the d-transition metals, whereas the lanthanides as a rule have the degree of oxidation 3; only in rare cases for these ions does it correspond to 2^+ and 4^+.

Crystal-Chemical Characteristics of Metals and Their Compounds

In solids material particles are arranged either in a definite order, forming a crystal structure, or at random, resulting in an amorphous condition. Metals are characterized by a regular arrangement of atoms and formation of crystalline forms.

The regularity of the internal structure of crystals is displayed in symmetry of their external bounding. During crystallization the growing crystal becomes covered by plane and straight edges. Faces, edges, and vertices are the bounding elements of the crystal and are interrelated by a dependence expressed by the well-known Euler-Descartes formula:

$$\text{Number of faces} + \text{Number of vertices} = \text{Number of edges} \pm 2.$$

According to present-day concepts, crystalline matter consists of atoms or ions arranged like points of the crystal lattice. The three-dimensional crystal lattice represents an infinite vec-

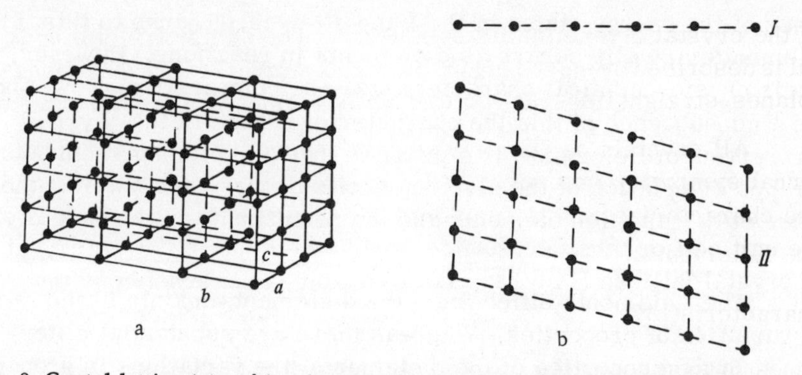

Fig. 3. Crystal lattice (a) and its elements (b): I) crystal lattice row; II) plane network.

torial structure in which the atoms or ions occupy definite geometric positions [28, 140].

Points arranged along one straight line make up a crytal lattice row (Fig. 3). The distance between two adjacent points is called the spacing of the row. In parallel directions the spacings are equal and in nonparallel they differ. Points arranged in one plane form a plane crystal lattice network consisting of parallelograms having adjacent sides entirely filling the space. The points of the crystal lattice are arranged at the vertices of the parallelograms. Three plane networks intersecting in space form a three-dimensional crystal lattice.* It consists of parallelepipeds at whose vertices are situated its points. These parallelepipeds are called repeating parallelepipeds, unit parallelepipeds, or unit cells.

The unit cell is characterized by the angles between the edges forming it (α, β, γ) and the ratio of edges (a:b:c).

The unit cell should have: a structure symmetry corresponding to that of the crystal; a maximum number of right angles; a maximum number of equal angles and equal edges; a minimum volume. The quantities α, β, γ, a, b, and c, are parameters of the unit cell.

The faces of real crystals are parallel to the plane networks of the lattice with the closest arrangement of the particles, and the edges are parallel to the lattice rows with the same arrangement of particles. The arrangement of the faces, edges, and vertices

*Usually the three-dimensional crystal lattice is called simply the crystal or space lattice.

of the crystal determine its symmetry. The symmetry of a crystal is described by means of auxiliary geometrical representations (planes, straight lines, and points) called elements of crystal symmetry.

All crystals can be divided with respect to external and internal symmetry into seven large groups called systems which are characterized by similarity of the angles between the edges of the unit cell or other crystallographic directions. The main types of crystal lattices [140] have the following structure symmetry characteristics:

Structure symmetry	Parameters of unit cell	
Cubic	$\alpha = \beta = \gamma = 90°$;	$a = b = c$
Tetragonal	$\alpha = \beta = \gamma = 90°$;	$a = b \neq c$
Rhombic	$\alpha = \beta = \gamma = 90°$;	$a \neq b \neq c$
Monoclinic	$\alpha = \gamma = 90°$; $\beta \neq 90°$;	$a \neq b \neq c$
Triclinic	$\alpha \neq \beta \neq \gamma \neq 90°$;	$a \neq b \neq c$
Trigonal	$\alpha = \beta = \gamma \neq 90°$;	$a = b = c$
Hexagonal	$\alpha = \beta = 90°$; $\gamma = 120°$;	$a = b \neq c$

The parameters of the unit cell determine its geometrical form, which depends not only on the magnitude of the angles but also on the ratio of edges. The cubic, tetragonal, and rhombic symmetries are characterized by right-angle unit cells. However, whereas in the case of the cubic system the cell has the form of a cube ($a = b = c$), in the case of the tetragonal system it has the form of a prism with a square cross section ($a = b$). For the rhombic system, which is characterized by inequality of the edges, the unit cell has the form of a match box.

The parallelepipeds of the monoclinic and triclinic systems of symmetry differ from the rhombic by oblique angles. In the monoclinic crystallographic system the angle between the edges a and c is obique and in the triclinic all angles are oblique. In the trigonal system of symmetry the unit cell is a rhombohedron which can be represented as a cube flattened or extended at opposite vertices. In this case all the edges remain equal ($a = b = c$) but the face acquires the form of a rhombus rather than of a square ($\alpha = \beta = \gamma \neq 90°$). The unit cell of the hexagonal system can be visualized as composed of three rhombic prisms; its cross section has the form of a regular hexagon.

Crystalline materials differ not only in the form of the unit cell but also in the arrangement of the particles in it. The number of possible space lattices with different arrangements of points is

14. These lattices are called Bravais space lattices (Fig. 4). There are four methods of formation of crystal lattices corresponding to E. S. Fedorov's four main types of structures:

primitive cubic cell (5) – the particles are arranged at the vertices of a cube;

body-centered cubic cell (8) – the particles are arranged at the vertices of a cube and at its center;

face-centered cubic cell (12) – the particles are arranged

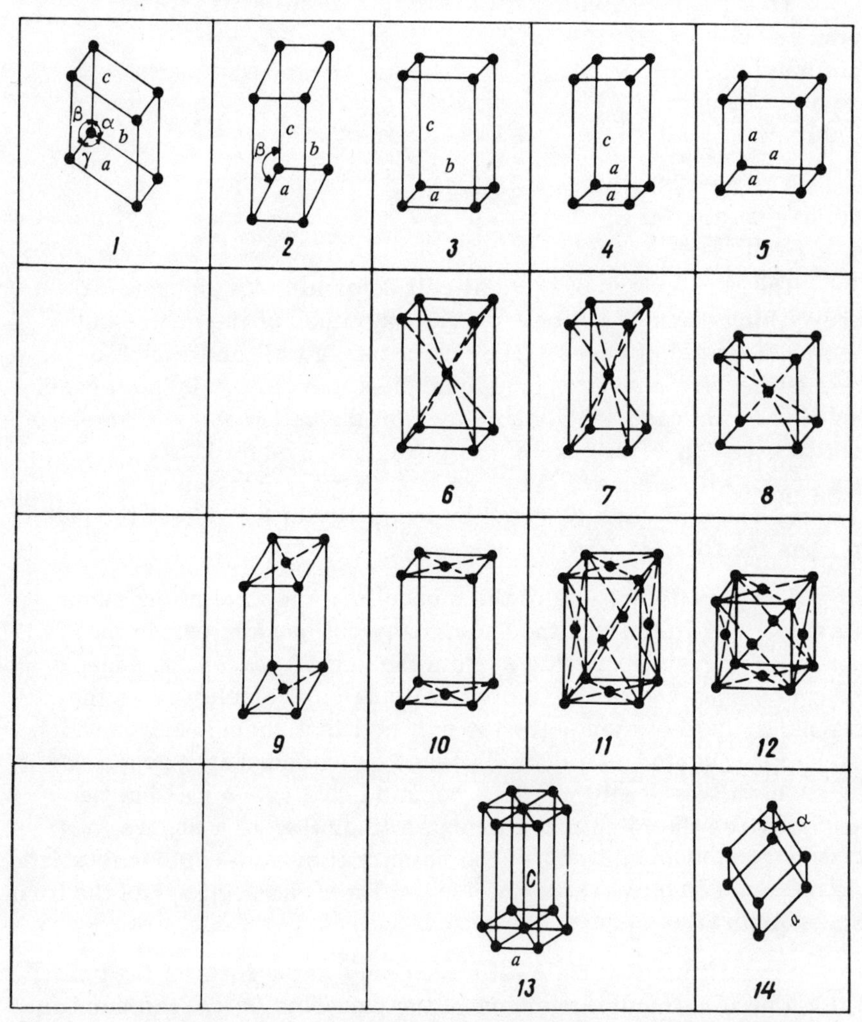

Fig. 4. Bravais lattices: 1-5) primitive; 6-8) body-centered; 9-10) base-centered; 11-12) face-centered; 13) hexagonal; 14) rhombohedral.

at the vertices of a cube and centers of the faces;

prismatic base-centered hexagonal cell (13) – the particles are arranged at the vertices of a prism and at the centers of the upper and lower bases.

The different electronic structures of the atoms and ions forming crystals determine their interaction, bonding forces, and relative arrangement in space, i.e., the crystal lattice structure.

To a first approximation the atoms or ions have a spherical form and a definite radius. The atomic or ionic radius is the minimum distance to which the center of the sphere of a given atom or ion approaches the surfaces of the nearest atom or ion (Table 2).

The metals of Groups IA and IIA of the periodic system have the largest atomic radii. Metals of Group IIIA (Y, Sc) and partially of Group IVA (Zr, Hf, Th) have smaller atomic radii. The atomic radii of heavy metals (In, Fe, Sn, Pb, Sb, Bi, Po) increase with increase of atomic number. The transition metals with an incomplete d shell have average atomic radii. The typical nonmetals with small atomic numbers (B, C, N, O, H) have the smallest atomic radii.

In the rows of the periodic system the ionic radii decrease with increase of their charges: Na^+, 0.98 Å, Mg^{2+}, 0.74 Å, Al^{3+}, 57 Å, Si^{4+}, 0.39 Å. In the groups of the periodic system the ionic radii increase with increase of the principal quantum number: Li^+, 0.68 Å, Na^+, 0.98 Å, K^+, 1.33 Å, Rb^+, 1.49 Å, Cs^+, 1.65 Å.

The lanthanide contraction is observed in the lanthanide group, i.e., a decrease of atomic radii with increase of the atomic number of the element: La^{3+}, 1.04 Å, Lu^{3+}, 0.80 Å. A decrease of the size of the atoms is observed also in other metals, for example, in the pairs silver–gold and zirconium–hafnium.

The ions from which crystals are built up can be represented as spheres touching each other. Under the effect of the electric charges of adjacent ions the spherical form of the ions is distorted, i.e., polarization occurs. The magnitude of polarization is directly proportional to the electric field strength.

One of the most important characteristics of the crystal structure is the coordination number, which corresponds to the number of nearest atoms and ions surrounding a given atom or ion

TABLE 2. Atomic and Ionic Radii of Some Elements, after Belov and Bokii [28]

Atomic number	Element	Radius, Å		Atomic number	Element	Radius, Å	
		atomic	ionic			atomic	ionic
3	Li	1,55	1+0.68 *	5	B	0,91	3+ (0.20)
11	Na	1,89	1+0.98	13	Al	1.43	3+0,57
19	K	2.36	1+1.33	31	Ga	1.69	3+0,62
37	Rb	2.48	1+1.49	49	In	1.66	3+0.92
							1+1.30
55	Cs	2,68	1+1.65	81	Tl	1.71	3+1.05
87	Fr	2,80	—				1+1.36
4	Be	1,13	2+0.34	6	C	0.77	4+(0.15)
12	Mg	1,60	2+0.74				4—(2,60)
20	Ca	1,97	2+1.04	14	Si	1.34	4+0,39
38	Sr	2,15	2+1.20	32	Ge	1.39 †	4+0,44
56	Ba	2,21	2+1.38	50	Sn	1,58	4+0,67
88	Ra	2,35	2+1.44	82	Pb	1.75	4+0,76
21	Sc	1,64	3+0.83	7	N	0,71	5+0,15
39	Y	1,81	3+0.97				3—1.48
57	La	1.87	3+1.04	15	P	1,3	5+0.35
			4+0.90	33	As	1.48	5+(0.47)
89	Ac	—	3+1.11	51	Sb	1,61	5+0.62
22	Ti	1.46	4+0,64	83	Bi	1,82	5+(0.74)
			2+0.78				
40	Zr	1.60	4+0.82	8	O	0,65	2—1,36
72	Hf	1.59	4+0.82	16	S	1.22	6+(0,29)
							2—1,82
23	V	1.34	5+0.4	46	Pd	1.37	4+0,64
			2+0.72				
41	Nb	1,45	5+0.66	78	Pt	1.37	4+0,64
73	Ta	1.46	5+(0.66)	58	Ce	1,83	5+0.88
							3+1.00
24	Cr	1.27	6+0.35	59	Pr	1.82	3+0.99
			3+0.64				
42	Mo	1.39	6+0.05	60	Nd	1.82	3+(0.98)
74	W	1.40	6+0.65	34	Be	1.93	6+0.35
			4+0.68				4+0.69
29	Cu	1,28	1+0.98	52	Te	2.11	6+(0.56)
							4+0,89
47	Ag	1,44	1+1.13	9	F	1,33	—
79	Au	1,44	1+(1.73)	25	Mn	1.30	7+(0,46)
30	Zn	1,39	2+0.83	43	Tc	1,36	—
48	Cd	1,56	2+0.99	61	Pm	—	3+0.97
80	Hg	1.60	2+1.12	62	Sm	—	3+0,97
75	Re	1.37	6+0.52	63	Eu	2,02	6+0,35
							4+0,85
26	Fe	1,26	3+0,67	64	Gd	1,79	3+0.94
44	Ru	1,34	4+0.62	65	Tb	1,77	3+0,89
76	Os	1.35	4+0.65	66	Dy	1.77	3+0.88
27	Co	1.25	3+0.64	67	Ho	1.76	3+0.86
45	Rh	1.34	4+0.65	68	Er	1.75	3+0.85
77	Ir	1.35	4+0.65	69	Tm	1.7	3+0,85
28	Ni	1.24	2+0.74	70	Yb	1.93	3+0,81
				71	Lu	1.74	3+0,80

*The plus and minus signs pertain respectively to positive and negative ions.

†According to Goldschmidt.

in a crystal. In ionic crystals the coordination number corresponds to the number of nearest ions of opposite sign.

For stability of the crystal structure it is necessary that the surfaces of the spheres surrounding the central ion touch the surface of the latter. If the central ion is so small that it freely moves in the interatomic space, the structure will be quite unstable; in such cases regrouping of ions leading to the formation of another, more stable structure with a lower coordination number is possible.

In 1922 Magnus determined the limits of stability of structures for different coordination numbers in relation to the radius ratios of cations (R_c) and anions (R_a) (Table 3).

The maximum possible coordination number, 12, is due to the closest packing of spheres of the same radius.

The concept of coordination polyhedra, i.e., imaginary polyhedra obtained by connecting the centers of nearest equivalent anions surrounding the cation, is closely associated with the coordination number (Figs. 5 and 6).

In the crystal structure of sodium chloride the Na ion has coordination number 6 – it is surrounded by six Cl ions, whose centers of gravity form an octahedral coordination polyhedron. In the structure of ZnS, where Zn has coordination number 4, the S ions are arranged at the vertices of a tetrahedron, etc.

A coordination polyhedron in the form of a cube corresponds to coordination number 8. For example, in the structure of CsCl the Cs ion is at the center of a primitive cube constructed on the Cl ions.

TABLE 3. Radius Ratios of Cations and Anions for Different Coordination Numbers (after Magnus and Goldschmidt)

R_c	R_a	Coordination number	Form of neighborhood
0—0.15	6.45— ∞	2	Dumbbell
0.15—0.22	4.45—6.45	3	Triangle
0.22—0.41	2.41—4.45	4	Tetrahedron
0.41—0.73	1.37—2.41	6	Octahedron
0.73—1.00	1—1.37	8	Cube
>1	—	12	Cubo-octahedron with 12 vertices

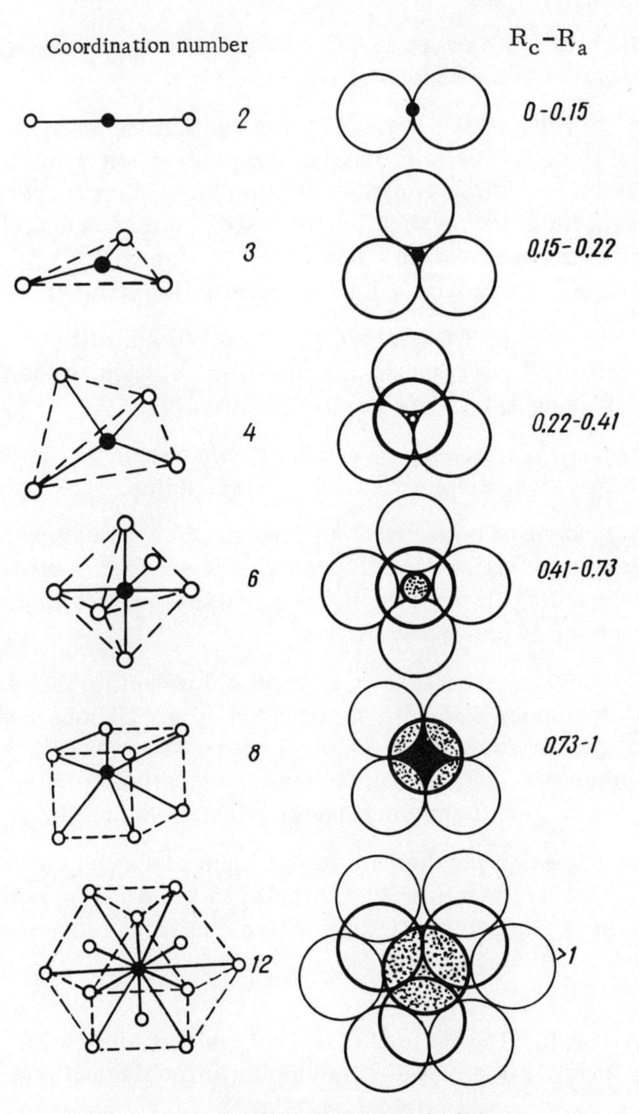

Fig. 5. Coordination polyhedra.

In crystals of various materials the bond between atoms and ions can be of four basic types: ionic, covalent, metallic, and molecular or van der Waals. Since the metallic type of bond corresponds to metallic crystals mainly, we will dwell on it in greater detail.

In 1900 Drude hypothesized that in metals the bond is realized by means of electrons removed from the metal atoms and forming

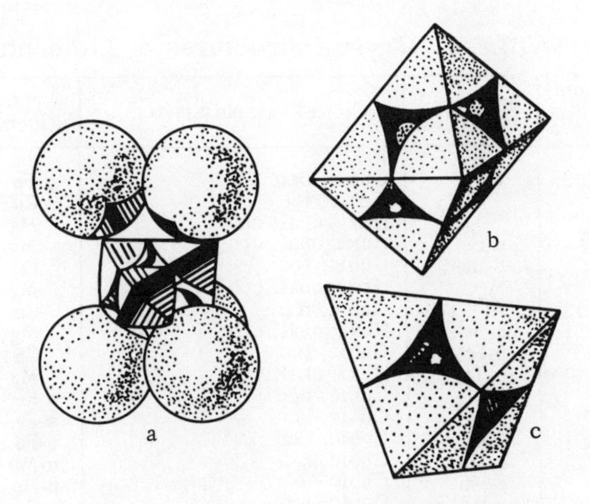

Fig. 6. Formation of coordination polyhedra (after Belov): a, b) Octahedral; c) tetrahedral voids.

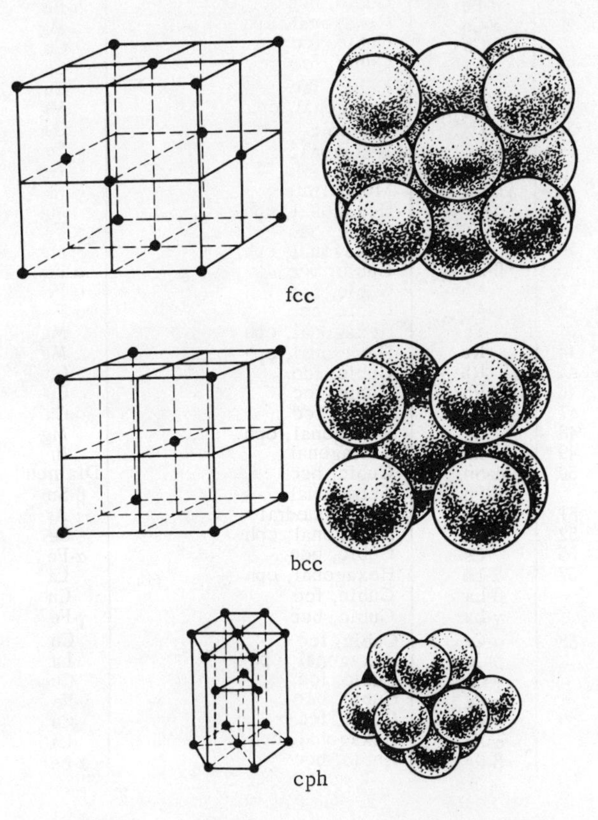

Fig. 7. Principal types of crystal lattices of metals.

TABLE 4. Crystal Structures of Elements

Atomic No.	Element	Crystallographic system	Type of structure
3	Li	Cubic, bcc	α-Fe
		Cubic, fcc	Cu
		Hexagonal, cph	Mg
4	α-Be	Hexagonal, cph	Mg
	β-Be	Cubic, fcc	Cu
12	Mg	Hexagonal, cph	Mg
13	Al	Cubic, fcc	Cu
21	α-Sc	Hexagonal, cph	Mg
	β-Sc	Cubic, fcc	Cu
22	α-Ti	Hexagonal, cph	Mg
	β-Ti	Cubic, bcc	α-Fe
23	V	Cubic, bcc	α-Fe
24	Cr	Cubic, bcc	α-Fe
25	α-Mn	Cubic	α-Mn
	β-Mn	Cubic	β-Mn
	γ-Mn	Cubic, fcc	Cu
	σ-Mn	Cubic, bcc	α-Fe
26	α-Fe	Cubic, bcc	α-Fe
	β-Fe	Cubic, bcc	α-Fe
	γ-Fe	Cubic, fcc	Cu
	σ-Fe	Cubic, bcc	α-Fe
27	α-Co	Hexagonal, cph	Mg
	β-Co	Cubic, fcc	Cu
28	Ni	Cubic, fcc	Cu
29	Cu	Cubic, fcc	Cu
30	Zn	Hexagonal, cph	Mg
31	Ga	Rhombic	Ga
34	Se	Hexagonal	Se
	α-Se	Monoclinic	—
	β-Se	Monoclinic	—
39	α-Y	Hexagonal, cph	Mg
	β-Y	Cubic, bcc	α-Fe
40	α-Zr	Hexagonal, cph	Mg
	β-Zr	Cubic, bcc	α-Fe
41	Nb	Cubic, bcc	α-Fe
42	Mo	Cubic, bcc	α-Fe
43	Tc	Hexagonal, cph	Mg
44	Ru	Hexagonal, cph	Mg
45	Rh	Cubic, fcc	Cu
46	Pb	Cubic, fcc	Cu
47	Ag	Cubic, fcc	Cu
48	Cd	Hexagonal, cph	Mg
49	In	Tetragonal	In
50	α-Sn	Cubic, bcc	Diamond
	β-Sn	Tetragonal	β-Sn
51	Sb	Rhombohedral	As
52	Te	Hexagonal, cph	Se
55	Cs	Cubic, bcc	α-Fe
57	α-La	Hexagonal, cph	La
	β-La	Cubic, fcc	Cu
	γ-La	Cubic, bcc	γ-Fe
58	α-Ce	Cubic, fcc	Cu
	β-Ce	Hexagonal, cph	La
	γ-Ce	Cubic, fcc	Cu
	σ-Ce	Cubic, bcc	γ-Fe
59	Pr	Cubic, fcc	Cu
	α-Pr	Hexagonal, cph	La
	β-Pr	Cubic, bcc	α-Fe

TABLE 4. Continued

Atomic No.	Element	Crystallographic system	Type of structure
60	α-Nd	Hexagonal, cph	La
	β-Nd	Cubic, bcc	α-Fe
62	α-Sm	Rhombohedral	Sm
	β-Sm	Cubic, bcc	α-Fe
63	Eu	Cubic, bcc	α-Fe
64	α-Gd	Hexagonal, cph	Mg
	β-Gd	Cubic, bcc	W
65	α-Tb	Hexagonal, cph	Mg
	Tb	Rhombic	—
66	Dy	Hexagonal, cph	Mg
		Rhombic	—
67	Ho	Hexagonal, cph	Mg
68	Er	Hexagonal, cph	Mg
69	Tu	Hexagonal, cph	Mg
70	α-Yb	Cubic, fcc	Cu
	β-Yb	Cubic, bcc	α-Fe
71	Lu	Hexagonal, cph	Mg
72	α-Hf	Hexagonal, cph	Mg
	β-Hf	Cubic, bcc	α-Fe
73	Ta	Cubic, bcc	α-Fe
74	α-W	Cubic, bcc	α-Fe
	β-W	Cubic, bcc	β-W
75	Re	Hexagonal, cph	Mg
76	Os	Hexagonal, cph	Mg
77	Ir	Cubic, fcc	Cu
78	Pt	Cubic, fcc	Cu
79	Au	Cubic, fcc	Cu
80	Hg	Rhombohedral	Hg
81	α-Tl	Hexagonal, cph	Mg
	β-Tl	Cubic, fcc	Cu
82	Pb	Cubic, fcc	Cu
83	Bi	Rhombohedral	As
84	α-Po	Cubic, fcc	Cu
	β-Po	Rhombohedral	—

an "electron gas" in which positively charged "ions" (cores of metal atoms) "float."

The bonding forces are evenly distributed in all directions (have spherical symmetry).

In metallic crystals the positive charges of the cores of the atoms and the negative charges of the free electrons are mutually balanced, and therefore intermetallic compounds can be formed easily without being restricted by the valency rule and obey only crystal-chemical laws.

The electrons occupy definite levels in isolated metal atoms. In crystals different regions of energies close in value, known as Brillouin zones, arise owing to the possibility of free migration of electrons from one atom to another.

The spherical symmetry of the bonding forces acting in metallic crystals determines the tendency towards the densest arrangement (closest packing) with the highest coordination number, 12, characteristic of their atoms. Three principal types of crystal structure are characteristic of metals: face-centered cubic (fcc) of the Cu type, close-packed hexagonal (cph) of the Mg type (both with coordination number 12), and body-centered cubic (bcc) of the α-Fe type with coordination number 8 (Fig. 7). Data on the character of the crystal structures of different elements are presented in Table 4 and Fig. 8.

The body-centered cubic lattice is characteristic of many metals: α-Fe, Cr, V, Mo, W, Nb, β-Ti, β-Zr, Ba, alkali metals, and an appreciable number of alloys.

Cubic close packing corresponds mainly to the structures of Cu, Au, Ag, Pb, Ni, Co, Pt, γ-Fe, Al, Sc, Ca, Sr, rare earth elements, beginning with Yb, and also a number of alloys.

A										B							
1	2	3	4	5	6	7	8			1	2	3	4	5	6	7	8
																H s^1	He s^2
Li s^1	Be s^2											B s^2p^1	C s^2p^2	N s^2p^3	O s^2p^4	F s^2p^5	Ne s^2p^6
Na s^1	Mg s^2											Al s^2p^1	Si s^2p^2	P s^2p^3	S s^2p^4	Cl s^2p^5	Ar s^2p^6
K s^1	Ca s^2	Sc d^1s^2	Ti d^2s^2	V d^3s^2	Cr d^5s^1	Mn d^5s^2	Fe d^6s^2	Co d^7s^2	Ni d^8s^2	Cu $d^{10}s^1$	Zn $d^{10}s^2$	Ga s^2p^1	Ge s^2p^2	As s^2p^3	Se s^2p^4	Br s^2p^5	Kr s^2p^6
Rb s^1	Sr s^2	Y d^1s^2	Zr d^2s^2	Nb d^4s^1	Mo d^5s^1	Tc d^5s^2	Ru d^7s^1	Rh d^8s^1	Pd $d^{10}s^0$	Ag $d^{10}s^1$	Cd $d^{10}s^2$	In s^2p^1	Sn s^2p^2	Sb s^2p^3	Te s^2p^4	I s^2p^5	Xe s^2p^6
Cs s^1	Ba s^2	La d^1s^2	Hf d^2s^2	Ta d^3s^2	W d^4s^2	Re d^5s^2	Os d^6s^2	Ir d^7s^2	Pt d^9s^1	Au $d^{10}s^1$	Hg $d^{10}s^2$	Tl s^2p^1	Pb s^2p^2	Bi s^2p^3	Po s^2p^4	At s^2p^5	Rn s^2p^6

Fig. 8. Crystal structures of chemical elements: 1) bcc; 2) fcc; 3) cph; 4) rhombohedral; 5) diamond type; 6) monoclinic; 7) tetragonal; 8) rhombic.

Hexagonal close packing is characteristic of less stable metals: Be, Mg, Zn, Cd, α-Ti, α-Zr, α-Hf.

The character of the crystal structure can change under the action of heating or other factors. For example, α-Ti has a hexagonal lattice, whereas β-Ti has a close-packed cubic lattice [185]; α-Be has a hexagonal and β-Be a cubic lattice.

Unlike metals, nonmetals are characterized by the existence of a covalent bond owing to the formation of electron pairs by interacting atoms. In this case stable electronic states are formed which approximate to the states of inert gases: s^2p^6 or less stable sp^3.

Both in the cubic and in the hexagonal close packings there are two types of voids between the spheres: Some are formed by four contacting spheres (tetrahedral voids) and others by six spheres (octahedral voids).

In ideal close packings of spheres the octahedral voids can contain atoms the ratio of whose radius to the radii of the spheres in close packing is greater than 0.414; for tetrahedral voids this ratio is 0.22. The number of octahedral voids (n) is equal to the number of spheres. The number of tetrahedral voids is equal to 2n.

The formation of stable electronic configurations of atoms is important in the evolution of physicochemical and other properties of metals. According to the concepts developed by G. V. Samsonov and his students [300, 302, 307, 308, 311, 312, etc.], during the formation of a crystalline solid from isolated atoms some electrons become localized near the cores of the atoms and some into a delocalized (collective) state. The localized electrons form in addition to very stable configurations, less stable and very unstable ones. In this case the statistical weight* of the energetically most stable electronic configurations exceeds the weight of the unstable ones. A certain number of stable configurations corresponds to each atom. A constant exchange occurs between the latter and the delocalized valence electrons, thanks to which a bond between electrons is realized.

*The statistical weight of stable electronic configurations of the localized valence electrons is the probability, expressed in percent, of the existence of a certain stable configuration of localized electrons in the spectrum of possible configurations.

In the formation of metallic crystals from isolated atoms the localized valence electrons form configurations differing in stability, among which the energetically most stable for d elements are d^0, d^5, d^{10} (approaching the stability of s^2p^6). The most stable is the d^5 configuration, whose electrons have the same spin directions. Stable configurations are covalent bond carriers [251, 378].

It was established by x-ray spectral analysis that, for zirconium, 2.6 electrons of the four ($d^2 + s^2$) valence electrons are localized and 1.4 are not localized. Of the six ($d^5 + s^1$) electrons of chromium 3.5 are localized; the delocalized part consists of 1.5 electrons (Korsunskii and Genkin, 1962, 1965). Recently G. V. Samsonov and his co-workers have calculated the statistical weight of localized and delocalized electrons for many metals (Table 5). Atoms having in an isolated state less than five electrons in the d shell form primarily stable d^0 and d^5 configurations. For atoms

TABLE 5. Statistical Weights of Stable Configurations of Transition Metals without Allowance for Intermediate Spectrum (after G. V. Samsonov)

Element	Electronic structure of valence subshells	Fraction of localized electrons, %	Fraction of delocalized electrons, %	Statistical weight of d^0 configurations, %	Statistical weight of d^5 configurations, %	Fraction of electrons of localized d^0 and d^5 configurations, %	Statistical weight of d^{10} configurations, %	Fraction of electrons of localized d^{10} configurations, %
Sc	$3d^1\ 4s^2$	30	70	82	18	30	0	0
Ti	$3d^2\ 4s^2$	94	6	57	43	94	0	0
V	$3d^2\ 4s^2$	63	37	37	63	63	0	0
Cr	$3d^5\ 4s^1$	58	42	27	73	58	0	0
Fe	$3d^6\ 4s^2$	76.3	23.7	0	78	49	22	27.3
Co	$3d^7\ 4s^2$	95	5	0	32	18	68	77
Ni	$3d^8\ 4s^2$	97	3	0	3	1.5	97	95.5
Y	$4d^1\ 5s^2$	37	63	73	22	37	0	0
Zr	$4d^2\ 5s^2$	65	35	48	52	65	0	0
Nb	$4d^3\ 5s^2$	78	22	24	76	78	0	0
Mo	$4d^5\ 5s^1$	73	27	12	88	73	0	0
Ru	$4d^7\ 5s^1$	87	13	0	60	37.5	40	50
Rh	$4d^8\ 5s^1$	97	3	0	25	14	75	83
Pd	$4d^{10}\ 5s^0$	98.8	1.2	0	2.4	1.2	97.6	97.6
La	$5d^1\ 6s^2$	38	62	74	26	38	0	0
Hf	$5d^2\ 6s^2$	69	31	45	55	69	0	0
Ta	$5d^3\ 6s^2$	80	20	16	84	80	0	0
W	$5d^4\ 6s^2$	74	26	6	94	74	0	0

with the same electronic configuration the energy stability of the valence d electrons increases with increase of the principal quantum number.

A similar regularity is observed among the f elements (lanthanides and actinides), for which the stable configurations are f^0, f^7, and f^{14}. The energy stability of these configurations also increases with increase of the principal quantum number of the f electrons.

The d^5 and d^{10} electronic configurations are the most stable for metal atoms having more than five and less than 10 electrons in the outermost shell. In this case the statistical weight of the d^{10} configurations increases as the number of electrons in the d shell approaches 10.

The maxima of the melting point correspond to the maximum statistical weight of the d^5 configurations. For each transition period the melting point of the metal rises with increasing statistical weight of the d^5 configurations and falls with increase of the d^0 and d^{10} states. The melting point of metals rises as the principal quantum number increases, i.e., with an increase of the energy stability of electronic configurations.

A rise in the melting point with the formation of stable f^7 electronic configurations and increase of the principal quantum number is characteristic of the group of f metals.

For metals of the copper subgroup (copper, silver, gold), having the electronic configuration $d^{10}s^1$, the occurrence of a high statistical weight of stable d^{10} and s^2 configurations at a small fraction of delocalized electrons is possible during the formation of crystals. However, owing to the presence in silver and gold of completely vacant 4f and 5f shells, in which partial electron transition is possible, disturbance of the formation of stable d^{10} and s^2 configurations occurs. Relatively stable d and f configurations are formed. Therefore, with increase in the principal quantum number and energy stability of both the d and f states a higher melting point (1060°C) is observed for gold than for silver (960°C). The statistical weight of delocalized electrons of these metals does not increase significantly.

For metals of the zinc subgroup (zinc, cadmium, mercury), having a $d^{10}s^2$ electronic configuration of isolated atoms, a high statistical weight of delocalized electrons arises during the formation of a metallic crystal due to excitation of s^2 configurations. This causes a fall in the melting points of these metals in comparison with the copper group metals. However, a partial transition of d electrons to the free f shell is possible both for silver and gold and for cadmium and mercury, which leads to a decrease of the statistical weight of the d^{10} configurations.

In the s metals group (alkali and alkaline earth metals of Groups I and II) electrophysical and other properties are evolved upon the formation of metallic crystals in the isolated state in connection with the formation of stable s^2 configurations. The atoms of alkali metals, having in the isolated state the s^1 configuration of the valence electrons, form stable s^2 configurations whose energy stability decreases with increase of the principal quantum number. This explains the decrease of the ionization potential from lithium to cesium (Li, 5.39, Na, 5.138, K, 4.339, Rb, 4.176, Cs, 3.893 eV).

The properties of beryllium, magnesium, and other alkaline earth metals having s^2 configurations are related with the possibility of $s \rightarrow p$ electron transitions and the formation of sp configurations characteristic of nonmetals. For beryllium the probability of $s \rightarrow p$ transitions is especially high. Here under conditions of excitation, for example, with rise in temperature, the $s^2 \rightleftharpoons sp$ equilibrium is apparently shifted to the right as a result of the considerable energy stability of the sp configurations under these conditions in comparison with the s^2 configurations characteristic of metal atoms.

The amphoteric properties of beryllium can be attributed to the formation of s^2 and sp configurations, which characterize metallic and nonmetallic properties.

The β modification of beryllium, characterized by the formation of sp^3 configurations, has a cubic crystal lattice, whereas the α modification, which is characterized by a predominance of s^2 states, has a hexagonal lattice (of lower symmetry).

The $s^2 \rightleftharpoons sp$ equilibrium shifts to the left for magnesium as a consequence of a decrease of energy stability of the sp configurations with increase of the principal quantum number, which is characteristic of nonmetals. This intensifies the metallic properties of magnesium in comparison with beryllium. Along with this, magnesium retains also certain nonmetallic properties due to the presence of sp electronic configurations.

Calcium and other alkaline earth metals are characterized by a decrease of energy stability of the stable s^2 configurations not only with an increase of the principal quantum number but also with the possibility of transfer of a part of the electrons to the free d level. In connection with this, on passing to calcium, strontium, barium, and radium the equilibrium shifts in the direction of $s \rightarrow d$ transitions, which gives rise to the metallic luster, the metallic thermal and electric conductivities, opaqueness, etc.

For crystals with structures of the same type and the same number of valence electrons an increase of interatomic distances leads to a fall of the melting and boiling points of metal and to an increase of the coefficients of thermal expansion and elasticity.

For metals with the same structures or with the same close packing and approximately equal interatomic distances the strength of the metallic bond increases with an increase of the number of valence electrons removed, which in turn raises the melting and boiling points and causes an increase of the hardness of the crystals and a decrease of the coefficient of expansion and elasticity.

The properties of metals can change as a result of interaction with atoms of other metals or nonmetals. In this case substitutional solid solutions (substitutional isomorphism) or interstitial solid solutions (interstitial isomorphism) can form.

Solid homogeneous mixtures of two or more components which do not interact with each other, form no compounds, and exhibit marked quantitative differences are called solid solutions.

Interstitial solid solutions – crystals of nonstoichiometric composition – are formed as a result of the arrangement of atoms of nonmetals in vacant sites of the structure, i.e, in the interatomic spaces. These compounds of metals with metals and also of metals with nonmetals have the general name of metallic compounds.

Many compounds of metals with metalloids with small atomic radii (carbon, nitrogen, hydrogen), called carbides, nitrides, and hydrides, belong to the aforementioned compounds, which are interstitial phases. Metallic properties, high melting point, and heat resistance are characteristic of these compounds; they constitute a large class of refractory compounds. According to Samsonov's classification [295] there are: refractory compounds of metals (metallic compounds proper or intermetallic compounds); metallike compounds of metals with metalloids (carbides, nitrides, phosphides, etc.); and metalloid–metalloid (nonmetallic compounds of boron with carbon, carbon with silicon, boron with nitrogen, etc.).

This classification is based on regularities of the change of character of the chemical bond in relation to the arrangement of the elements in the periodic system.

Compounds of the first class (aluminides, antimonides, beryllides, etc.) have a relatively simple structure with a predominance of the metallic bond, and usually form a cubic or hexagonal lattice with close packing of the atoms, more rarely a tetragonal lattice.

The type of chemical bond and crystal structure of the compounds belonging to the second class are determined by the share of participation of the metals and metalloids in the formation of the compounds.

Compounds of the second class are characterized by heterodesmic structures, in which the relationships of various chemical bond forces can change depending on the electronic structure of the components and their quantitative composition.

Upon interaction of nonmetals compounds are formed with a covalent bond and properties characteristic of nonmetals (boron carbide, silicon carbide, boron nitride, silicon nitride, etc.).

Compounds of metals with metal analogs differing insignificantly in metallochemical properties (for example, Au, Cu) are the least stable chemically.

In compounds formed as a result of the interaction of atoms with pronounced metallochemical characteristics there occurs a gradual change from the metallic type of bond to the covalent and even to the ionic. The properties of the compounds change accordingly.

The properties and structure of borides and other refractory compounds have been studied by many authors [289-301, 305, 306, etc.]. Since boron has a large atomic radius (r_B = 0.87 Å) which prevents the arrangement of its atoms in the interatomic spaces, it was at first considered that borides could not be regarded as interstitial phases. However, later investigations proved such a possibility for borides of metals of Groups IV–VI. These atoms can be arranged in the lattice of the metal both in an isolated state and in the form of chains, skeletons and networks, which is determined by the number of boron atoms in the compound. In the case of high boron contents (MeB_4, MeB_6, MeB_{12}) the formation even of three-dimensional skeletons is possible.

The metallic bond is the main type of bond between the metal and boron atoms in borides. However, with an increase of boron content the share of the ionic bond increases. At the same time the covalent bond is retained between the boron atoms [136, 291, 295].

The level of the metallic bond in borides can be estimated quantitatively with the formula proposed by Samsonov, namely, $1/Nn$, where N is the principal quantum number of the incomplete d shell and n is the number of electrons at the d level. The greater the $1/Nn$ ratio, the more boron electrons at the d level of the atoms of transition metals and the more pronounced the metallic character of the bond.

This regularity pertains not only to borides but also to other refractory compounds. However, in the opinion of Kornilov [136], this ratio becomes disturbed with an increase of the content of the atoms of metalloids which interact with one another on the basis of the covalent bond.

Borides are characterized by the formation primarily of hexagonal, rhombic, and tetragonal crystal lattices—with a less perfect symmetry. This is related with the large atomic radius of boron, the introduction of which into the crystal lattice of the metal leads to its loosening.

Simple carbides also belong to typical interstitial phases with close-packed cubic or hexagonal lattices. Carbides are characterized by metallic properties (luster, high electric conductivity, etc.).

Unlike the compounds described, silicides (compounds of metallic elements with silicon) do not form interstitial phases, but belong primarily to substitutional phases.

The silicon atoms form ordered structures (chains, skeletons, layers) or are in an isolated state. However, unlike borides, silicides have paired silicon atoms. Covalent bonds form between the silicon atoms in compounds. In addition, silicides have metallic bonds also, which is indicated by their physicochemical properties. In most cases silicides are characterized by the formation of hexagonal, tetragonal, and rhombic structures, i.e., crystal lattices with a weak degree of symmetry are characteristic.

Nitrides, like other refractory compounds, are typical interstitial phases. The atomic radius of nitrogen ($r_N \simeq 0.71$ Å) is smaller than the atomic radius of carbon ($r_C = 0.76$ Å). Nitrides, in common with carbides, have metallic properties, which is apparently due to the possibility of transition of part of the electrons to the incomplete d level of transition metals as a consequence of a decrease of energy stability upon an increase of the interatomic distances in interstitial phases. The formation of cubic and hexagonal close packings is characteristic of these compounds.

Thus the main groups of refractory compounds constituting interstitial phases are characterized by considerable similarity of structure and peculiarities of internal bond formation behavior, which is responsible for their similar physicochemical properties.

A metallic type of bond is formed in compounds of metals with oxygen, sulfur, phosphorus, arsenic, and other metalloids, which gradually, as the number of atoms of the metalloids increases, changes to a covalent and then ionic bond.

Examination of the crystal structures formed by refractory compounds shows that the majority of them are characterized by a NaCl-type of lattice (face-centered cubic) or by the more close-packed lattice of the nickel arsenide NiAs. Weakly symmetric structures are less frequently encountered among them. The crystal-chemical features of these compounds, like the crystal structures of metals, are due to their electronic structure [136, 137, 140, 295]. The presence in the metal atom of valence electrons in a quantity sufficient for filling the outermost p shell of the metalloid to the stable s^2p^6 configuration is the condition for the occurrence of the NaCl- and NiAs-type structures with the coordination number 6.

Upon formation of compounds of d-transition metals with nonmetals there occurs on one hand the release of part of the valence electrons with the formation of stable states characteristic of d-transition metals and, on the other, the build-up of the sp orbits to the stable sp^3 and s^2p^6 states. For example, upon formation of compounds of d-transition metals with boron having the s^2p configuration, it is probable that the boron atom will tend to acquire the high statistical weight of the stable sp^2, and then the sp^3 configuration. This was proved by comparing the structural elements of boron atoms in crystal lattices of boride phases.

The statistical weight of sp^3 configurations increases as a consequence of the transition of part of the electrons from the metallic component to the nonmetallic and depends on the electronic structure of the metal atoms and their ability to give up valence electrons.

In the case of formation of borides of Group IV transition metals, the metals of this group (Ti, Zr, Hf), having the electronic configuration d^2s^2, can transfer part of the collective electrons, with the formation of stable sp^3 configurations of the electrons of the boron atoms. In metals of Groups V-VI the possibility of transferring part of the delocalized electrons for the formation of stable sp^3 configurations of the boron atom is limited owing to the increase of the statistical weight of the stable d^5 configurations. Therefore, the hardness, melting point, and other properties of borides of Group V are much lower than for borides of Group IV metals. In view of this borides of Group VI have a maximum statistical weight of the sp^3 states. As a consequence of this they also have minimum hardness.

Carbides are characterized by build-up of the s^2p^2 configuration of the carbon atom to sp^3 owing to the addition of part of the nonlocalized electron of metals. Metals of Group IV, having a low statistical weight of the stable electronic d^5 states, are able to transfer a considerable part of the electrons, which is related to the high statistical weight of the stable sp^3 configurations of the carbon atoms. Titanium, having a minimum statistical weight of stable d configurations, has the maximum ability to give up electrons. In zirconium and hafnium, which are characterized by a large statistical weight of the d^5 states, the possibility of giving up the delocalized part of the valence electrons manifests itself more weakly, as a consequence of which the melting point, hardness, and other physical properties decrease in carbides of these metals. In connection with the foregoing, in carbides of Groups IV and VI metals a lesser ability to form stable sp^3 configurations is noted, which gives rise to an even greater decrease of their hardness and other properties.

In chromium carbides there is a strong tendency toward formation of stable sp^3 configurations of carbon atoms, since several carbon atoms figure in their composition. Therefore, chromium carbides have a high statistical weight of sp^3 configurations and hence a high hardness, strength, etc.

In connection with the decrease of the energy stability of nonmetals with an increase of the principal quantum number, considerably more nonlocalized electrons are required in the formation of silicides as a result of interaction of silicon atoms (s^2p^2) and metal atoms for the formation of stable sp^3 states than in the case of formation of borides and carbides.

In contrast to this, in the formation of nitrides (compounds of metals with nitrogen) we should expect the appearance of a high statistical weight of the sp^3 states, which is confirmed by analysis of certain physical properties.

Other compounds of metals with nonmetals can be considered in the same manner from the viewpoint of electronic structure. A covalent bond forms upon interaction between nonmetals, as a consequence of which these compounds have properties characteristic of nonmetals: weak electric conductivity, brittleness, hardness, etc.

Complex Compounds of Metals and the Formation of a Coordination Bond

There is still another form of strong exchange interaction, called a coordination bond, which is important for understanding processes occurring in inorganic and organic compounds [13, 14, 86, 452].

This type of bond occurs upon formation of complex compounds of metals, which include many catalysts, dyes, etc. Chlorophyll participating in plant photosynthesis, hemoglobin permitting respiratory processes, and many enzymes are also among these compounds. Metal ions entering the blood form complex compounds with active groups of proteins and nucleic acids. Antidote treatment in metal poisoning is also related with the use of complexing agents. Metal carbonyls, many of which are presently used as catalysts and for obtaining high-purity metal powders, are complex compounds.

Compounds whose molecule or ion contains a central atom or ion (usually of a metal) surrounded by other ions or molecules, called addends or ligands, are called coordination or complex compounds.

The most common types of ligands are monatomic and polyatomic negative ions and neutral polar molecules. The last include primarily molecules having one or several unshared pairs of electrons, for example, NH_3, H_2O, and CO.

The coordination bond in complex compounds is regarded in most cases as covalent. However, it differs from the typical covalent bond in that, when it ruptures, both electrons participating in the formation of the paired bond pass to one of the interacting atoms. The structure and properties of complex compounds have been examined thoroughly in [13, 14, 24, 86].

In most cases complex compounds have a tetrahedral structure with the coordination number 4 or an octahedral structure with the coordination number 6 (Fig. 9).

The formation of complexes of d-transition metals with ligands is closely related with the electron distribution functions of the atom. A one-electron wave function of the atom represents the product of the angular and radial distribution functions. The radial distribution function is determined by the distance of the electron from the

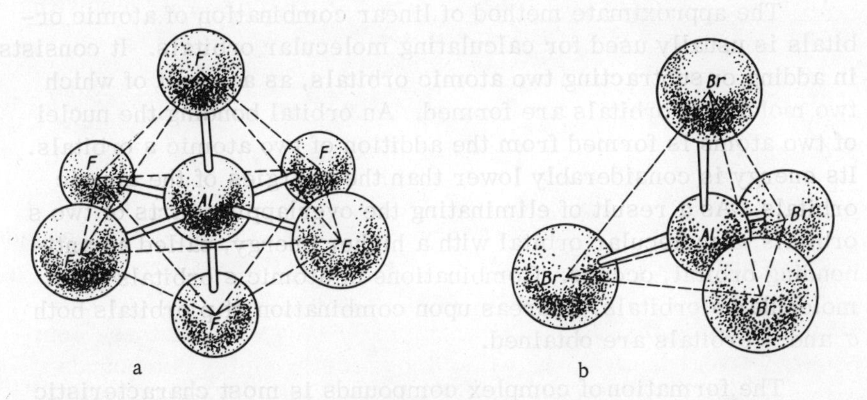

Fig. 9. Crystal structure of coordination compounds: a) octahedral (AlF_6^{3-}); b) tetrahedral ($AlBr_4^-$).

nucleus, and the other function is determined by the angular coordinates of the electron. The angular distribution functions, regardless of the type of atom being considered, are related only with the distribution of the electrons among the orbitals. Only one angular distribution in the form of a sphere is characteristic of s electrons, three of sp electrons, and five of d electrons. The electrons of the p orbital are distributed in the form of a dumbbell oriented along one of the three Cartesian coordinate axes. The p_x, p_y, and p_z orbitals are oriented along the x, y, and z axes, respectively (Fig. 10).

Four of the five d orbitals have the form of a cloverleaf and one the form of a dumbbell with a ring around the center. Three cloverleaf orbitals, d_{xy}, d_{xz}, and d_{yz}, are oriented, respectively, in planes xy, xz, and yz, so that they are situated between two axes; the fourth, $d_{x^2-y^2}$, is oriented in the xy plane along the x and y axes. The single orbital oriented along the z axis has the form of a dumbbell.

At present there is no one theory for explaining the formation of coordination bonds. The electrostatic crystal-field theory, valence bond method, and theory of molecular orbitals are used widely for this purpose.

The most popular at present is the theory of molecular orbitals, which takes into account the ionic and covalent type of bond in coordination compounds. According to this theory, electrons in molecules are distributed in orbitals (as in atoms).

The approximate method of linear combination of atomic orbitals is usually used for calculating molecular orbitals. It consists in adding or subtracting two atomic orbitals, as a result of which two molecular orbitals are formed. An orbital bonding the nuclei of two atoms is formed from the addition of two atomic s orbitals. Its energy is considerably lower than the energies of the atomic orbitals. As a result of eliminating the overlapping parts of two s orbitals, a molecular orbital with a higher energy, called an antibonding orbital, occurs. Combinations of atomic s orbitals form molecular σ orbitals, whereas upon combination of p orbitals both σ and π orbitals are obtained.

The formation of complex compounds is most characteristic of atoms of transition metals (Ti, V, Cr, Mn, Fe, Co, Ni, etc.); in these cases the close values of the energies of the ns, np, and (n^{-1}) d levels of the metal atom allow all nine valence (sp^3d^5) orbitals

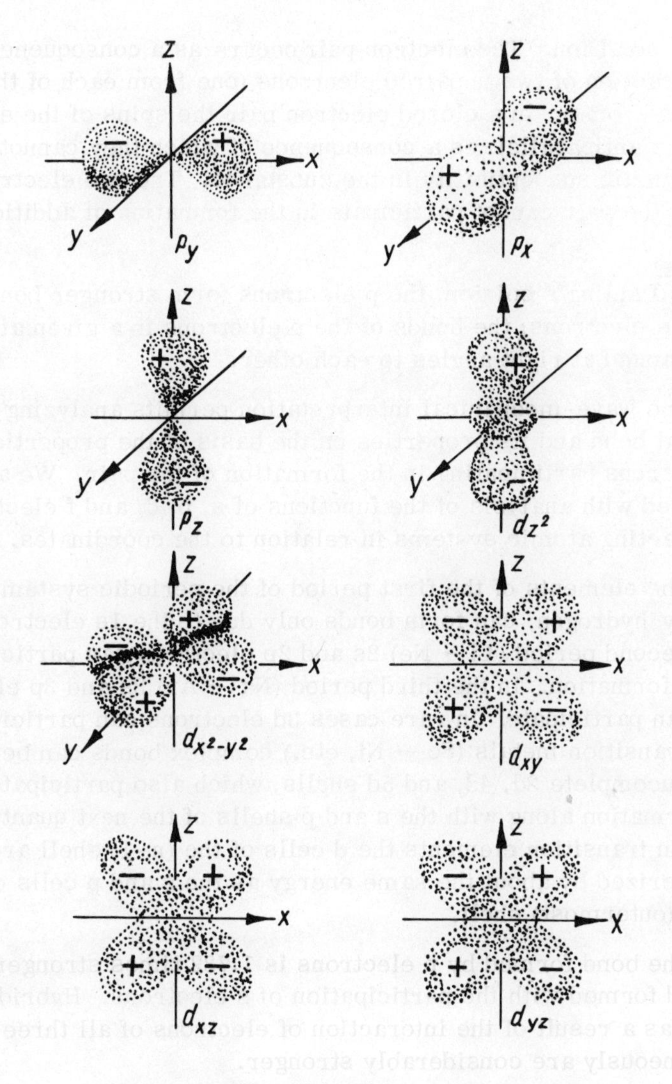

Fig. 10. Spatial configuration of p and d orbitals.

to participate in chemical-bond formation. This leads to the occurrence of nine bonding molecular orbitals of the entire complex, which, according to the Pauli principle, can be occupied by 18 electrons.

According to current concepts of quantum mechanics, complex compounds are formed as a result of the interaction of the electron pair of the ligand being coordinated and the electron shell of the central ion. The electron pair occurs as a consequence of the interaction of two unpaired electrons (one from each of the interacting atoms). In a closed electron pair the spins of the electrons are antiparallel, as a consequence of which they cannot cause paramagnetic susceptibility in the substance. The two electrons forming the pair cannot participate in the formation of additional pairs.

In Pauling's opinion, the p electrons form stronger bonds than do s electrons; the bonds of the p electrons in a given atom are arranged at right angles to each other.

The wave-mechanical interpretation permits analyzing the chemical bond and its properties on the basis of the properties of the electrons participating in the formation of the pair. We are concerned with analysis of the functions of s, p, d, and f electrons of interacting atomic systems in relation to the coordinates.

The elements of the first period of the periodic system, particularly hydrogen, can form bonds only due to the 1s electron. In the second period (Li → Ne) 2s and 2p electrons can participate in bond formation. In the third period (Na → Ar) 3s and 3p electrons can participate. In rare cases 3d electrons can participate. In the transition metals (Sc → Ni, etc.) complex bonds can be formed due to incomplete 3d, 4d, and 5d shells, which also participate in bond formation along with the s and p shells of the next quantum level. In transition elements the d cells of the (n−1) shell are characterized by about the same energy as the s and p cells of the second (outermost) shell.

The bond formed by p electrons is 1.732 times stronger than the bond formed with the participation of s electrons. Hybrid bonds arising as a result of the interaction of electrons of all three shells simultaneously are considerably stronger.

The strongest hybrid bonds are formed with the participation of the d cells. Therefore, elements with unfilled d cells are considered the most typical complexing agents: Ni ($3d^8 4s^2$), Co ($3d^7 4s^2$), Cr ($3d^5 4s^1$), Pd ($d^{10} 5s^0$), Pt ($5d^9 6s^1$), Ir ($4d^2 5s^2$).

In accordance with Hund's rule, the nickel atom in the ground state has two unpaired d electrons, the cobalt atom three, the chromium atom six (five 3d and one 4s), the palladium atom none, the platinum atom two (one 5d and one 6s), and the iridium atom three. The appearance of other covalence values is related with change of the atoms to an excited state. Taking the foregoing into account, we arrive at the conclusion that complexing with palladium is possible only in the case of an excited state with transition to the $4d^95s$ (more probable) or $4d^95p$ (less probable) state. Such regroupings also affect the strength of complexing in systems of nickel, cobalt, and other elements.

The d cells of the penultimate shell and the s or p cells of the outermost shell participate in the formation of the primary bonds in complex compounds. Since the two individual bonds formed in this case are less strong than the hydrid bonds dsp^2 or d^2p^2, another two "auxiliary" valences appear due to vacant cells of the outer shell. That only two p cells are used in this case is explained by the effect of energy factors. According to calculations, hybrid dsp^2 bonds are energetically more advantageous than dsp^3 bonds. In this way, the formation of complex compounds can be visualized from the viewpoint of covalent bond theory.

The change of these compounds to derivatives of tetravalent elements is linked with the opening of still another pair of electrons occupying the d cell. For platinum this process consists in opening yet another pair in the 5d cell with transition of one of the electrons to the 6p level and subsequent hybridization of the bonds (d^2sp^3). This process is considerably hindered in the case of palladium and especially nickel.

In the case of cobalt and iridium the manifestation of the chief valence, equal to three, is possibly due to the formation of pairs by way of the two unpaired p electrons and one s electron of the outermost layer. The three outer p cells with subsequent hybridization (d^2sp^3) are used for the manifestation of auxiliary valence. As regards the chromium atom, in the manifestation of its hexavalence it uses all six unpaired electrons for the formation of simple or double bonds. Only part of the unpaired electrons is used in the manifestation of lower values of the chief valence (for example, equal to three). As judged by the properties of the complex derivatives of trivalent chromium, the formation of d^2sp^3 bonds is probable in this case.

In addition to the aforesaid we must point out that if three d eigenfunctions participate in the formation of the bond, the tetrahedral arrangement is more stable.

This interpretation of the mechanism of the coordination of complex compounds is due largely to Pauling [256]. The main role is assigned to the characteristics of the structure of the electron shells of the complexing atoms. It is in fact the properties of the complexing atom and the value of the coordination number that determine geometric configuration. At the same time, such an approach is most valid in the case of typical complexing agents for which the value of the coordination number and geometric configuration do not depend on the nature of the ligands that can form dsp^2 or d^2p^2 bonds. Ions with filled d shells (for example, Zn^{2+}, Cd^{2+}, Hg^{2+}, Cu^+, Ag^+, Au^+) can form bonds only with the participation of s and p electrons, i.e., they should not give rise to complexes with coordination number 4.

In other cases the stability of the chemical bond is affected by the nature of the attached ligands. This problem and the entire theory of the structure of complex compounds are expounded in detail in [86].

The data given in this chapter on the electronic structure and crystallochemical properties of crystals of various metallic elements and their compounds indicate a close correlation between the electronic and crystal structures of substances and the evolution of their physical properties.

There is no doubt that the character of the electronic and crystal structures of elements governs the possibility of the formation of metallic disintegration and condensation aerosols and determines the shape of the particles, electric charge, air-suspension time, and their toxic properties.

CHAPTER II

Conditions of Formation of Metallic Aerosols in the Air of Industrial Rooms in Powder Metallurgy and Allied Plants

Aerosols of metals are one of the principal unfavorable factors in the production of ferrous and nonferrous metals, in the foundry industry, at plants of the machine-building and electrical-engineering industry, powder metallurgy plants, in arc welding operations, etc. Metallic aerosols are found also at other industrial installations, in particular at plants of the porcelain and faience, glass, lacquer and paint, and light industries, where metal oxides are used as pigments and coloring materials. During packing, grinding, preparation of paints, their application to the surfaces of articles, and in other processes such substances enter the air of the working zone and can have a deleterious effect on the workers.

Plants at which metallic dust is the main harmful factor can be divided into two main groups. The first includes plants where the formation of dust is linked with pyrometallurgical processes (nonferrous and ferrous metallurgy plants, arc welding shops, etc.).

High temperatures produce metal vapors, which cool on entering a zone with a lower temperature, forming condensation aerosols. Being in an excited state due to high temperatures, the metal atoms actively combine with atmospheric oxygen, forming oxides. Thus, metallic condensation aerosols are oxides.

The second group includes industries in which cold processing of metals and metal powders predominates. The processes of dust formation at such plants are related with comminution, screening, transporting, and unloading pulverized metal products.

These industries include, in particular, powder metallurgy plants. They produce metal powders (pure metal powders and powders of various metallic compounds) and sintered articles for various purposes (electrical contacts, ferrites, magnets, machine parts, antifriction and friction units, thermocouples, and oxidation-resistant sheaths). The starting materials for the manufacture of sintered articles are primarily powders of iron, copper, titanium, zirconium, hafnium, niobium, chromium, molybdenum, tungsten, and other metals, and also of refractory metallic compounds (borides, carbides, silicides, tellurides, etc.). These substances can arrive at powder metallurgy plants in a finished form. Some of them (e.g., iron powders and powders of refractory compounds) are manufactured in special shops.

Industrial Hygiene in the Production of Refractory Compounds

Depending on the scale of manufacture refractory compounds, they can be produced in a special plant or in a separate shop of the plant. At the present level of development of the industry, refractory compounds as a rule are produced in separate shops of powder metallurgy plants or in special shops of other plants.

The principle of the chemical reaction for producing the main types of refractory compounds – carbides, borides, silicides – is analogous to the reduction of metals. The production of other compounds – nitrides, sulfides – is based on the interaction of solid and gaseous substances by reactive diffusion.

In production of refractory compounds and articles from them, the starting materials (metallic oxides, metallic silicon, boric anhydride, sulfur) are rough crushed in jaw or hammer crushers, ground in ball or other mills, and blended in various types of mixers. In certain cases the metal powders are first reduced. The prepared mix is charged, usually manually, into graphite boats and placed in continuous electric furnaces (most often of the Tammann type) for chemical reaction. The material obtained is crushed, ground in ball mills, screened, and again subjected to thermal reactions. The finished product is then unloaded from the boats, crushed, ground, screened, and packed in cans for shipping to other plants.

If such refractory compounds are to be used at the same plant, they are sent to the milling and preparing department of the powder metallurgy shop.

Finely divided powders of oxides or reduced rare and rare earth metals and nonmetals (metallic silicon, boron, carbon black, sulfur, phosphorus, and also gaseous substances – nitrogen, hydrogen, hydrogen sulfide, hydrogen selenide, hydrogen telluride), etc., are used most often for the synthesis of refractory compounds. In boride production the dust of metallic components (chromium, titanium, zirconium) and also boric oxide, boron, carbon black, and carbon monoxide can enter the air of the working zone. In carbide production the liberation of dust of metals and nonmetals (boron, silicon) and carbon black is possible. The synthesis of metallic silicides is accompanied by the formation of metallic aerosols, metallic silicon, and, possibly, silicon dioxide and carbon monoxide. In the case of sulfide production the air can contain dust of the starting metallic compounds, sulfur, hydrogen sulfide, and other sulfur compounds. The production of selenides and tellurides can be attended by the evolution of hydrogen selenide, hydrogen telluride, and powdered metals.

Dust is released into the air of the working zone during crushing of the materials, weighing of the components of the mix manually, loading and unloading the milling and mixing equipment, screening the materials, charging them into boats, packing the finished product, and transporting the starting pulverized materials and finished powdered compounds.

TABLE 6. Pollution of the Working Zone Atmosphere in the Production of Refractory Chromium Compounds

Operation	Dust concentration, mg/m³			Quantity of dust particles (%) of size:		
	minimun	maximun	mean	<1 μ	1–5 μ	>5 μ
Weighing of chromic oxide	10.6	19.6	12.3	16	63	21
Loading and unloading charge from milling equipment	260.0	640.0	482.5	14.5	63.0	22.5
Loading chromium carbide charge into graphite cartridges	24.0	40.0	33.7	—	—	—
Screening chromium carbide and packing	120.0	199.6	144.9	7.5	63	29.5

Maximum dust pollution of the air is usually found in the milling and preparing department, especially in the absence of effective local exhaust ventilation and effective enclosure of dust-producing units.

In chromium carbide production the maximum dust content during weighing of the chromium oxides in the mix was found to be 19.6 mg/m^3 (Table 6). During the loading of the charge manually into the drums of the roller tables the dust content in the breathing zone of the operator was 260-640 mg/m^3 (average 482.5).

In the heat-treatment department with manual loading of the charge into graphite boats or cartridges the dust content was 24-40 mg/m^3 (average 33.7).

A high level of dust pollution is found also during screening and packing of the finished product in cans, when the chromium carbide content of the air is 120 mg/m^3 and higher. Approximately the same level of dust pollution of the air was noted in the production of zirconium boride and molybdenum disilicide.

During manual charging of zirconium carbide into boats the dust content reached 190 mg/m^3. Similar results were obtained by Mezentseva [201] for the manufacture of tungsten carbide. According to these data, the dust content during the preparation of the charge and loading and unloading of the ball mills was 20-83 mg/m^3. During loading of the charge into boats the dust content was 10-46 mg/m^3; during loading and unloading of the ball mills, 33.3 mg/m^3 and higher. During loading of the charge into the boats and unloading of tungsten carbide from them the dust content reached 17.5-47 mg/m^3, and during screening of tungsten and its carbide on vibrating sieves, 8.3-42.2 mg/m^3. According to the data of [202], in the production of titanium carbide the dust content in the carbonization zones (preliminary stages of the production process) was within 20-42 mg/m^3 and during screening of titanium carbide 20.3-40.2 mg/m^3.

During weighing and loading of chromic oxide into ball mills 79% of the particles measured up to 5 μ, during screening of chromium carbide 63-70%, and during packing of the finished carbide in cans 68-70%.

The dust in the production of tungsten carbide was very fine. Particles measuring up to 4 μ amounted to 72-80%.

It should be noted that in the milling and preparing department the workers are exposed to the combined effect of dust and medium-frequency noise whose level during operation of the milling and mixing equipment can reach 80 dB.

In the heat-treatment department not only dust can be produced but also convective and, to a lesser extent, radiant heat generated well as gaseous products (nitrogen oxides, carbon monoxide, etc.) formed as a result of nitriding, carbidization, and other processes related with the interaction of gaseous and solid components.

In the production of carbides the content of carbon monoxide in the working zone near the charging ports of the electric furnaces reached 4.0-10.0 mg/m^3 even during operation of a local exhaust ventilation. In certain cases the carbon monoxide content here exceeded the maximum allowable concentration by a factor of two or three; the content of nitrogen oxides in the zone of the furnace operators during the production of nitrides reached 6.2 mg/m^3. The air temperature here reached 28-30°C at a relative humidity of 50-59%.

The occurrence of an unfavorable microclimate is largely a result of inadequate thermal insulation of the outside surfaces of the furnaces. As investigations showed, the temperature of different areas of the furnaces varies from 31 to 122°C, appreciably exceeding the permitted level (45°C). In this case there is a direct relation between the temperature of the outside surfaces and that prevailing inside the furnace. Thus, if at a temperature of 100°C inside the furnace the surface of the front wall had a temperature of 33.2°C, then at 780°C the temperature was 73°C, at 1500-1600°C it was 73.4-93°C, and at 1800-2000°C the wall heated up to 107-122°C. The surface of the cooler was heated less, since its temperature did not exceed 36°C, i.e., was within the allowable limits.

The surfaces of structures close to the furnaces also had a high temperature: In particular, the temperature of the surfaces of the exhaust hoods located over the charging ports of furnaces was 41-57°C.

In a study of the working conditions during the synthesis of boron nitride, which is carried out in the same shop as the synthesis of other refractory compounds of transition metals, the highest dust pollution of the air was observed during preparation of the charge. During manual loading of the charge components

TABLE 7. Dust Contents of the Working Zone in the Production of Boron Nitride

Operation	Dust concentration, mg/m^3		
	minimum	maximum	mean
Loading of boric oxide and carbon black into ball mill and unloading of charge	92.3	352.5	243.2
Loading of charge into crucibles	6.4	18.3	12.5
Knocking out of cake	5.0	32.0	15.5
Loading of boats with boron nitride	32.3	48.0	41.3
Unloading boron nitride from boats	56.7	79.2	71.4
Screening of boron nitride after first stage of nitriding	25.0	130.0	68.6
Rubbing of boron nitride through screens	13.3	32.2	25.4
Screening of finished product without ventilation	22.2	126,0	52.1
Packing of boron nitride in cans in open zones	33.0	112.0	73.4

into the ball mill (without local extraction) the level of airborne dust reached 92.3–352.5 mg/m^3 (Table 7). Less dust pollution of the air was noted during other production operations, which were performed under covers equipped with local exhaust fans. Even these, however, proved to be insufficient. High dust pollution (tens and hundreds of milligrams per cubic meter was found in the air of the working zone during screening and packing of the finished boron nitride, which were done manually in areas not equipped with covers and local exhaust fans.

Dust of nonmetallic substances constituting the most harmful component of the industrial atmosphere is formed in shops producing boron and silicon carbides.

TABLE 8. Dust Pollution of Air in Working Zone during Production of Boron Carbide

Operation	Dust concentration, mg/m^3		
	minimum	maximum	mean
Crushing of starting materials	330.0	5280.0	1694,0
Preparation and release of charge	377.0	1930,0	1077.0
Discharge of finished product from furnace	530.0	4156.0	1898.2
Fractionation of boron carbide on screens			
Door of screen housing closed	0.5	6.0	–
Door of screen housing open	19	24.7	22.2

In the production of boron carbide a high airborne dust content was observed also in the milling and preparing department when charging the raw material (boric acid and coal tar pitch) into the crushing and milling equipment, during mixing of the charge, and loading of the charge into the electric arc furnaces. The highest level of dust-laden air was noted during dismantling of the furnace walls and unloading of the finished boron carbide (Table 8).

In the department where boron carbide is subjected to fine wet grinding the air as a rule is not laden with dust. Work in this department is usually accompanied by medium- and high-frequency noise within 80-85 dB.

An effect on the workers handling the finished product was observed only in shop areas for drying the finished boron carbide, its classification, and during packing. Screening of the pulverized materials was done on screens equipped with a cover and fans. The level of airborne dust during screening of boron carbide powders with the door tightly closed was 0.5-6.0 mg/m^3 and with the door open 19-24.7 mg/m^3 (average 22.2 mg/m^3).

In the manufacture of boron carbide the workers operating the electric arc furnaces are exposed, apart from dust, to radiant and convective heat, low-frequency noise, and a high air content of hydrocarbons, carbon monoxide, and other gaseous substances formed during the interaction of the starting ingredients. Similar data for working conditions in the production of boron carbide are described in [51].

A high dust content was noted also in the production of silicon carbide. In the first stages of the production process the workers may have to breathe air contaminated with the starting components of the charge: quartz sand, crushed coke, sodium chloride, and sawdust (during their mixing and charging into the electric furnaces). The dust of the finished silicon carbide can enter the breathing zone of the workers during its unloading from the electric furnaces, classifying, milling, screening, and packing of the pulverized material. The dust pollution during unloading and classifying of silicon carbide is negligible, since these processes are usually performed in open areas and the crystalline mass of silicon carbide is initially moistened for rapid cooling. A different picture is observed in the shop areas during screening and during packing of the material in kraft-pulp bags. The content of silicon

TABLE 9. Silicon Carbide Dust Content in Air during Screening and Packaging

Sampling site, operation	Dust concentration, mg/m^3		
	minimum	maximum	mean
Hopper feeding pulverized material to central screen	78.0	260.0	129.0
Separator	55.0	276.6	145.8
Start of screen	126.0	231.3	162.0
End of screen	34.6	50.0	39.7
Screening of coarse-grained powder (M-20 and M-28)			
door of housing open	31.0	54.0	46.7
door of housing closed	4.0	10.6	8.0
Screening of fine-grained grinding powder			
door of housing open	158.0	820.0	453.8
door of housing closed	17.8	29.0	21.4
Unloading powders from hopper and packing in bags	44.0	110.0	69.0
Cleaning up the room	31.6	168.8	84.0

carbide dust in various screening areas amounts to tens and hundreds of milligrams per cubic meter (Table 9).

The level of airborne dust depends directly on the tightness of the doors of the screen housing. With tightly closed doors the content of silicon carbide dust did not exceed 4-10.6 mg/m^3 and with open doors 31-54 mg/m^3.

Dust pollution is also governed by the degree of comminution of the material: As the particle size of powders decreases the content of dust increases, irrespective of the screening processes being performed. While in the screening of coarse-grained powders (M-20 and M-28) the dust content reached an average of 46.1 mg/m^3, in the department for screening fine grinding powders it reached tens and hundreds of milligrams per 1 m^3 in various areas. With closed doors of the screen housing the dust content in the air dropped to 17.8-29.0 mg/m^3, but with open doors it reached 158.0-820.0 mg/m^3.

High dust pollution of working zones was noted in places where the powder was fed from the hopper to the screens (78-260 mg/m^3) and also where it was removed by hand from under the screens (44-110 mg/m^3, average 69.0). The air dust content due to the bagging of the material was slightly lower, 40-75 mg/m^3

(average 56.8). Considerable dust formation was observed during dry cleaning of the room.

The main cause of high dust formation in all shop areas where refractory compounds are produced is the inadequate mechanization and sealing of the operations of loading and unloading the dust-forming materials from the crushing and milling equipment and their transportation, often by hand, in unsealed containers. As a rule, nonstandard equipment is used, the design of which does not provide adequate sealing and local extraction to prevent the liberation of dust from the materials into the air of the working zone.

Hygienic Characteristics of Working Conditions in the Production of Iron Powders

Iron powder is one of the most widespread materials used for the fabrication of sintered friction and bearing articles in machine building. Of special interest for industry is the production of carbonyl iron powders, which are used for the manufacture of magnetic powder inductance coils employed in a variety of components of electronic equipment and as fillers of magnetic powder clutches and brakes [235, 324, 335, 344]. In this case a necessary condition is the isolation of the powder particles from each other, since contact between carbonyl powder particles markedly lowers the electromagnetic properties of the articles.

To meet the needs of powder metallurgy and other branches of technology, the production of iron powders by the reduction and carbonyl methods has been organized at some plants.

Production of Iron Powders by the Reduction Method. There are several methods of producing iron powders. Methods based on the reduction of iron oxides (mill scale, ore, and other industrial wastes) are the most common at Soviet plants. Lamp black, converted natural gas, charcoal, and coal tar pitch are used as reducing agents.

Mill scale arriving at a plant is unloaded from railroad cars into storage rooms, and then sent to special furnaces (ZhASh) for drying and later for grinding. The pulverized scale is mixed with carbon black and sent to hoppers for loading into special containers – trays or buckets. The charge thus packed is placed in horizontal or vertical continuous furnaces for reduction. As a result of the interaction of hydrogen, converted natural gas, and partially of carbon (black) with the oxygen of iron oxides (scale) under high-temperature conditions, reduced iron, in the form of sponge, is obtained. The sponge is first pulverized by means of hammer, roller, or jaw crushers and then sent for fine grinding and fractionation. The finished powder is packed in cans and hermetically sealed.

The main industrial hazard in shops producing iron powder by reduction in the first stages of the process is the dust of the scale and carbon black, and at the end of the production process the finely divided iron powders. When charcoal and coal tar pitch are used as reducing agents the dust of these materials can affect the workers in the first stages of production. In addition to dust the reduction furnace operators can be exposed to high temperature, infrared radiation, and a high content of carbon monoxide in the air.

A high air temperature adversely affects operators near the drying furnaces of the starting materials (the ZhASh furnaces)

The working conditions of workers in the iron powder production shops are determined by the degree of sophistication of the equipment used.

In a shop with little mechanization of production processes the dust level reached 235-597 mg/m^3 during manual unloading and loading of the scale in the milling equipment for preliminary drying and during mixing of the scale with carbon black. During filling of trays with the charge (manually) in open areas not equipped with exhaust ventilation the air dust content was 100.0-200.0 mg/m^3. In the milling department, in areas equipped with exhaust ventilation, the dust content was at the level of 7-12 mg/m^3 during milling of the iron sponge, screening of the iron powders on vibrating screens, and their packing in cans [30, 32, 33, 129].

The air temperature in the working zones near the charging ports of the reduction furnaces in summer (outside air temperature 22°C) rose to 32-33°C at a relative humidity of 45%. A considerable role in the formation of an unfavorable microclimate is played by thermal radiation originating mainly from the heated surfaces of the furnaces, whose temperature in the shop under consideration reached 60-80°C.

The carbon monoxide content at the furnace operators' position near the charging ports of the furnaces reached the maximum allowable concentrations or only slightly exceeded the permissible level, which apparently can be attributed to effective operation of the exhaust system, whose receiving hood was placed over the charging area of the furnace.

During a study of the working conditions in other plants [106, 141] with an improved organization of the production process, escape of dust into the air in specially constructed shops, where the main production operations were mechanized, was noted mainly at places of reloading of materials, during loading of the charge into trays, and during unloading and packing of the finished powder. A high air dust content, reaching tens and hundreds of milligrams per cubic meter, was detected during charging of the scale by the open method into hoppers feeding the scale-drying furnaces by means of bucket cranes, during delivery of carbon black manually to the mixing equipment, and during loading of the trays with the charge. According to the data of [106], during manual charging of black into the mixing barrels the dust level in the working zone reached 200-385 mg/m^3; during unloading of the charge from the mixing barrel 70-80 mg/m^3; and during its charging into the trays 20-100 mg/m^3. Dust formation was observed in places where the sponge was knocked out of the trays (50-90 mg/m^3) and crushed in hammer mills (60-160 mg/m^3), and also during classification on screens (40-50 mg/m^3) and during bagging (90-100 mg/m^3).

The cause of dust formation in these shop areas is inadequate sealing of the dust-producing equipment and lack or inefficiency of exhaust fans where dust is produced.

A higher dust content of the air was noted in the production of iron powders in a shop having fewer mechanized operations and numerous additional production lines for crushing charcoal and its briquetting with coal tar pitch and scale (iron oxides). The dust content in the breathing zone of the workers reached tens of milligrams per cubic meter. Thus, with this method of reduction an additional harmful agent can enter the air of the working zones, namely, coal tar pitch dust.

In the heat-treatment department, where such substances are used as reducing agents and periodic shaft furnaces are used for reducing scale, carbon monoxide, and various hydrocarbons, possibly including 3,4-benzopyrene can be liberated in the breathing zone of the workers.

According to experimental data, the carbon monoxide content near the charging furnaces in most cases exceeded the maximum allowable concentration (MAC) by a factor of five.

A high carbon monoxide content was noted during operation of horizontal continuous furnaces, but its content in the air of the working zone as a rule was within 30-40 mg/m^3, i.e., exceeded the MAC by a factor of 1.5-2. The untoward effect of these deleterious factors on furnace operators can be intensified through the action of radiant and convective heat.

The air temperature during operation of the horizontal furnace was on the average 26-33°C and during charging of the shaft furnaces increased to 35°C. The temperature of the outside surfaces of the furnaces and adjacent structures in some instances reached 90-100°C. The intensity of radiation near the open hatches of the furnaces reached 2.5-3 cal/cm^2 per minute.

Taking into account the possible effect of the aforementioned factors on workers, we can say that the reduction of scale in horizontal continuous furnaces with the use of converted natural gas and carbon black as reducing agents is more favorable in a hygienic respect.

Production of Carbonyl Iron Powders. Carbonyl powders of grade R-10, high purity iron powders of grade V-13, and carbonyl iron powders whose particles are coated with insulating SiO_2 films, bakelite lacquers, etc., are usually produced by the carbonyl method.

Carbonyl iron powders are usually produced in a special sealed apparatus [18, 335, 336] whose main elements are a vaporizer, a decomposer, and a vibrating filter. In the tank of the vaporizer liquid iron pentacarbonyl is thermally converted to vapor, which, on entering the decomposer under high temperature and pressure conditions, and also under the action of carbon monoxide, is decomposed with the formation of iron particles and carbon monoxide. Because of the difference of temperatures corresponding to the three zones of the decomposer, the iron particles are in constant motion. As a consequence of this, the carbonyl iron particles in cross section have a variable density and a characteristic onion-skin structure. The density of the layers of carbonyl iron powders deposited in the hot areas with a high content of carbon monoxide is less than that of the layers deposited in the low-temperature zone.

Primary separation of the iron powders take place in the decomposer. The large particles settle out from a suspended state and are deposited in the receiving hopper under the decomposer, and the small particles are carried out by the gas flow to the vibrating filter, and are also deposited in a receiving hopper, which is periodically unloaded. Before starting the process the apparatus is first purged with ammonia and then filled with carbon monoxide. During loading of the hoppers and cleaning of the apparatus carbon monoxide and pentacarbonyl vapor can enter the air of the working zone if the safety rules are violated. Considerable quantities of finely divided carbonyl powder (grade R-10) can enter the breathing zone of the workers. The airtightness of the device can be disturbed during sampling.

In the production of high-purity ions (V-3 and others) the iron powder is subjected to fractionation, sometimes with the use of a Ganel classifier, and to additional thermal reduction in Tammann-type electric furnaces with subsequent milling, screening, and airtight packing.

As investigations of the working conditions showed, a high dust content in the air is an unfavorable factor in this industry. High dust concentrations are observed during manual unloading of the hoppers of the decomposer and vibrating filter.

In the case of ineffective exhaust ventilation the quantity of dust in the breathing zone of the workers during handling of the materials reached appreciable concentrations (Table 10). An even

TABLE 10. Dust Content in Air of Working Zone during Production of Iron Powders by the Carbonyl Method

Operation	Dust concentration, mg/m^3		
	minimum	maximum	mean
Unloading of carbonyl iron powder from hopper of decomposer and vibrating filter	6.0	320.0	123.3
Loading of container with carbonyl iron powder	0	5.6	3.2
Opening of container with iron powder after heat treatment	430.0	612.0	536.0
Crushing of reduced iron sponge	0	0.4	0.3
Screening of V-3 powder	35.0	194.0	89.1
Grinding of high-purity iron in ball mill	0	1.0	0.3
Screening of dried SiO_2-coated carbonyl iron powder	54.0	96.0	73.3

higher dust formation was noted during cleaning of the lines of the apparatus by means of compressed air or manually.

At further production stages the air dust content was noticeably lower. During classification of the powders and their loading into containers, accomplished with exhaust ventilation, the dust content in the breathing zone did not exceed 10 mg/m^3.

Insignificant dust formation was observed also during crushing of the reduced carbonyl iron, but during manual screening the dust content increased to 35.0-194.0 mg/m^3 (average 89.0).

During unloading of the carbonyl powders from the hoppers of the decomposer and vibrating filter, and also during opening and cleaning of the apparatus, the workers can be exposed to appreciable concentrations of carbon monoxide.

It was established that the carbon monoxide content in the air of the working zone in the course of these operations in some instances exceeded the established MAC (20 mg/m^3) by a factor of 10-40.

Carbon monoxide, vapors, and liquid iron pentacarbonyl can also reach the air of the working zone in the case of poor sealing of the lines of the apparatus, at places where samples are taken, and also in the event of an accident. With inadequate insulation of the surface of the decomposer the ambient air temperature can rise.

Thus, release of finished metal powders, vapors, liquid iron pentacarbonyl, and carbon monoxide into the air of the working zone is possible during a number of production operations. During further heat treatment of the powders the furnace workers are exposed to high temperature and infrared radiation, as well as to dust.

The iron powder dust formed upon reduction of scale is extremely fine (54-80% of the particles measure under 5 μ). The dust of carbonyl iron is even finer containing 80-90% of particles measuring less than 2 μ.

In the production of iron powders whose particles are coated with a SiO_2 film, dust having a high fibrogenic activity enters the air of the working zone in the screening, packing, and other shop areas. A distinguishing feature of carbonyl production is the serious danger to health from exposure to high concentrations of carbon monoxide and highly toxic iron pentacarbonyl.

Conditions of Formation of Metallic Aerosols during the Manufacture of Sintered Articles

The manufacture of powder metallurgy articles calls for a preliminary reduction and roasting of the starting powdered metallic and nonmetallic charge components in electric furnaces, milling, weighing in prescribed proportions, and thorough mixing, and blending. The prepared charge is sent for pressing of semifinished products, which is accomplished by automatic, semiautomatic, hydraulic, or other presses [350].

The pressed semifinished articles are loaded into graphite boats or crucibles and packed with powder – pulverized alumina and charcoal with an admixture of other components preventing welding together of the articles under the action of high temperature during sintering. The articles thus packed are charged into continuous furnaces (of the Tammann type), Globar batch furnaces, or others. Exposed to high temperature (up to 3000°C), the charge components sinter together and the articles acquire the strength and other qualities stipulated by the specifications. Articles with a high accuracy of shape after sintering undergo sizing, i.e., final pressing, and are subjected to further heat treatment.

After sintering some sintered parts are machined. Structural parts, manufactured mainly from iron base materials, are sometimes subjected to sulfurization, impregnation with mineral

TABLE 11. Dust Content in Working Zone during Manufacture of Sintered Articles from Iron-Base Powder

Operation	Dust concentration, mg/m³		
	minimum	maximum	mean
Screening of iron powders	7.3	12.0	10.2
Screening of graphite powder	10.0	27.0	15.8
Mixing of iron powder with graphite and plasticizer, loading and unloading of ball mills	46.5	72.0	59.5
Pressing of iron–graphite bushings on semiautomatic presses	0	1.3	0.32
Manual proportioning of iron–graphite mixture	4.0	29.5	19.5
Pressing of iron–graphite articles on hydraulic presses (at press operator's position)	0.1	5.0	2.7
Loading of articles into boats with packing powder	33.3	200.0	97.2
Separating of articles from packing material by screening	46.5	133.2	81.5

oils, or quenching. The finished articles are inspected, packed, and stored.

Powder metallurgy plants have the following departments: milling and mixing, or composition preparation, pressing, heat-treatment, furnace, workshop, inspection, and packing. The main untoward factor of the industrial environment at all stages of manufacture of sintered articles is a high dust level of the air of the working zone. Maximum dust production is noted in the milling and mixing departments, where jaw, hammer, and roller crushers and ball, eddy, rod, vibrating, and colloidal mills are usually used as crushing and milling equipment.

Mixing of the charge components is accomplished in continuous mixers (vibrating, circulating, and centrifugal) or with periodic loading (cone, screw, centrifugal, batch).

The materials are screened in centrifugal air separators or by mechanical screening on vibrating and rubbing screens. In some cases screening is done manually.

The dust content at the operators' positions in the milling and mixing departments in the case of poorly mechanized production

TABLE 12. Dust Content at Operators' Positions during Manufacture of Copper-Base Sintered Articles

Operation	Dust concentration, mg/m³		
	minimum	maximum	mean
Loading of cupric oxide into boats	31.1	246.6	188.8
Loading of bronze–graphite mixture into milling-and-mixing equipment (manual)	13.3	1332	93.9
Manual loading of vibrating screen and screening of material (without cover)	60	324	166.6
Screening of bronze-graphite mixture on vibrating screen under cover	10.6	42.0	11.3
Loading of bronze–graphite mixture into container	311.8	548.0	347.5
Charging of hopper of semiautomatic press	45.9	67.3	58.9
Pressing of articles			
on semiautomatic press	6.6	272.0	69.5
manually	73	100.5	86.3
Sizing of bronze-graphite bushings	0	5.8	1.7
Unloading of tungsten–copper pole pieces from crucibles with packing material	5.3	143.1	52.5
Dressing of articles on emery stone	25	107	79.3
Wet grinding of articles	10	23.3	16.6
Dry grinding of articles	233	802	584.6

TABLE 13. Dust Content at Operators' Positions during Manufacture of Silver Contacts

Operation	Dust concentration, mg/m³		
	minimum	maximum	mean
Loading and unloading of silver–cadmium mixture from ball mill	24.3	27.4	25.8
Loading and unloading of silver–nickel–graphite mixture from hexahedral mixer	223.15	664.5	439.0
Screening of silver–cadmium mixture on vibrating screens	66.5	152.1	91.5
Screening of graphite on vibrating screens	66.9	97.2	83.5
Pressing of silver contacts on semiautomatic and hydraulic presses	3.3	48.7	13.6
Unloading of articles from boats and separating from packing material after sintering	85.5	223.5	163.4
Sizing of silver contacts	0	9.2	8.3

reaches high values (Tables 11-16). The dust content in the air of the working zone increases considerably where too little or no exhaust ventilation is provided at the main places of dust release. In the manufacture of iron-graphite-base articles the maximum dust content in the mixing department (averaging 59.5 mg/m^3) was noted during mixing of the constituents and loading and unloading the material from the ball mills; high dust contents were observed during the manufacture of copper contacts and other copper-base articles; the dust content in the breathing zone of the workers while loading cupric oxide into boats in open areas not equipped with local exhaust ventilation reached 138.8 mg/m^3 on the average. A slightly

TABLE 14. Dust Content in Air of Working Zone during Manufacture of Nonferrous Contacts

Operation	Dust concentration, mg/m³		
	minimum	maximum	mean
Manual weighing of charge components	21.0	28.0	23.6
Milling, loading, and unloading of tungsten–nickel contact charge from ball mills	11.5	45.3	37.3
Pressing of tungsten–nickel articles on semiautomatic presses	11.3	44	23.3
Pressing of tunsten–nickel contacts on 160-ton Orenburg hydraulic presses	0	5.0	2.7
Manual proportioning of charge	15.3	28.6	19.2
Loading of articles into boats and their packing	30.6	45.8	41.3
Unloading of tungsten–nickel contacts and their separation from packing material	113.7	631.5	326.5

TABLE 15. Dust Content in Air of Working Zone during Manufacture of Ferrobarium Magnets

Operation	Dust concentration, mg/m³		
	minimum	maximum	mean
Weighing of charge components, mixing with kaolin and barium	86.8	162.2	142.0
Loading of ferrobarium magnet charge into vibrating mill	35.4	47.1	41.3
Unloading of ferrobarium charge from vibrating mill	24.5	65.7	60.2
Loading of ferrobarium charge into batch mixer, mixing with ethanol, and unloading	31.3	44.5	35.6
Loading of ferrobarium into disintegrator and rubbing	21.5	42.5	32.5
Pressing of ferrobarium magnets on semiautomatic presses	3.5	8.1	6.0
Manual weight proportioning of charge	10.6	41.5	26.0
Pressing of ferrobarium magnets on 160-ton Orenburg presses	6	20.0	6.2
Loading of ferrobarium charge into crucibles for roasting	40.0	89.0	65.0
Unloading of ferrobarium charge after heat treatment	60.8	209.0	134.9

TABLE 16. Dust Content in Air of Working Zone during Manufacture of Nonferrous Magnets (of Alnico and Magnico types)

Operation	Dust concentration, mg/m³		
	minimum	maximum	mean
Crushing of ferroaluminum master alloy in jaw crusher	131.0	234.0	193.0
Screening of nonferrous magnet charge on vibrating screen	22.6	252.0	186.0
Loading and unloading of nonferrous magnet charge from ball mill	13.3	65.6	48.8
Pressing of nonferrous magnets (Alnico and others) on semiatomatic presses	0	1.3	0.6
Pouring of nonferrous magnet charge into hopper of semiautomatic presses	57.5	121.0	89.3
Loading of nonferrous magnets into boats and their packing with carbon powder containing iron, nickel, aluminum, and titanium			
Local ventilation not operating	46.6	153.2	96.2
Local ventilation operating	13.3	41.35	27.3
Unloading of articles from boats (local ventilation operating)	14.5	19.5	16.7
Dry polishing of nonferrous aluminum–nickel–cobalt magnets (Alnico) on flat polisher	64.5	242.0	146.7

lower dust content was noted during loading and unloading of the vibrating mills and eddy mixer, and also when screening the materials. The air dust content dropped considerably (on the average to 11.3 mg/m^3) when the materials (bronze-graphite mixture) were screened on vibrating screens under a cover (see Table 12). Approximately the same dust pollution of the air was found during similar operations in the manufacture of silver contacts and nonferrous magnets.

A lower air dust content (about 37.3 mg/m^3 on the average) was observed during the same operations in the manufacture of tungsten-nickel contacts and ferrobarium magnets, although the organization of labor at these operators' positions in the mixing department was much the same. This permits the assumption that the decrease of the dust level in this case is related with the higher specific gravity of the main charge components (tungsten, nickel, and iron powders), whereas in the manufacture of copper- and silver-base articles and nonferrous magnets the charges contained lighter powdered materials.

The dust produced during the manufacture of various sintered articles is very fine. The majority of dust particles measure 2-3μ (Table 17). Even finer dust was noted in the manufacture of copper-base articles and nonferrous magnets.

The composition of the dust entering the air of the working zone is determined by the composition and purpose of the sintered articles. In the manufacture of nonferrous electrical contacts, tungsten, copper, and nickel dust was found in the working zone; in the production of bronze-graphite bushings, there was copper, tin and graphite dust. In the production of iron-base articles, iron and graphite dust is liberated into the air of the working zone, and in the manufacture of ferrites the dust of iron and magnesium ferrites, $MnCO_3$, $CuCO_3$, $BaCO_3$, kaolin, and other components, the composition of which is determined by the nature of the ferrites.

In ferrobarium magnet manufacture the dusts of iron oxides, barium nitrate, and kaolin were liberated, and in the manufacture of nonferrous magnets the dust of iron-aluminum, iron–zirconium, and iron–titanium alloys, and also copper and cobalt.

The production of another type of magnets (Alni) was accompanied by the release of the dust of iron, nickel, copper, and iron–aluminum, iron–titanium, and iron–aluminum–zirconium alloys.

TABLE 17. Particle Size Analysis of Dust at Operators' Positions during Manufacture of Sintered Articles, %

Operation	Size of dust particles, μ						
	<1	1-2	2-3	3-4	4-5	5	6-7
Unloading of ferronickel master alloy from mill	34.8	32.4	13.8	11.2	6.2	1.6	–
Screening of ferronickel master alloy on vibrating screen (middle of shop)	41	23	4.8	7	21.2	2.4	0.6
Pressing of aluminum-nickel-cobalt magnets	50.8	14.8	11.2	12.2	1.2	9.2	0.6
Pressing of articles (middle of pressing department)	58	16	4.4	5.8	8.2	5.4	2.2
Loading of ferrobarium charge into mixer	37.6	26.8	18.4	8.2	7.6	1.4	–
Pressing of ferrobarium magnets	29	34.4	15.8	10	6.6	3.4	0.8
Pressing of tungsten-copper articles	18.4	15.6	20.8	7.8	16.6	16.6	4.2
Unloading of copper-graphite mixture from mixers	30.4	30.8	14.2	8	10.8	4	1.8
Milling of ferroaluminum master alloy in muller	24.6	20	10.6	11.8	30	3	–
Proportioning of iron-graphite mixture	16.6	20.6	20.8	29.4	8.2	4.4	–
Pressing of ferrobarium magnets	33.4	27.4	22.2	8.2	4.6	2.8	1.4
Screening of copper-graphite mixture of vibrating screens	58	27.8	8.2	4	1.4	0.6	–
Milling of ferroaluminum master alloy in muller	18	24.2	18.6	14.4	15.2	6.4	3.2
Pressing of copper-graphite articles (PV-474 press)	57.8	16.6	8.4	4.8	5.6	4.8	2
Pressing of magnesium-nickel-cobalt magnets (Magnico)	50.8	14.8	11.2	12.2	7.2	3.2	0.6
Pressing of tungsten-copper contacts	34.6	32.6	13.8	11.2	6.2	1.6	–
Pressing of bronze-graphite articles (PV-474 press)	30.4	31.0	14.2	9	10.0	4	1.4
Loading and unloading of ferrobarium charge	18	36	26	12	5	1	2
Loading of nonferrous magnets into boats	15	25	38	10	12	–	–
Unloading of nonferrous magnets after sintering	10	24	36	8	15	2	5
Loading of iron-graphite articles into boats with packing material	13	19	42	7	14	3	2
Unloading bronze-graphite articles from boats	14	28	32	15	9	2	–
Separating from packing material	18	30	24	18	10	–	–

During the production of bronze–graphite bushings the air can be polluted with copper dust with an admixture of graphite and tin; a chemical analysis of this dust showed that it contained 67–98% copper.

The dust produced during the fabrication of articles from iron-base powders (bushings, etc.) can contain iron, graphite, and other components. The composition of the dust of the milling and mixing departments is not limited to the list of substances given.

The powders of titanium, zirconium, hafnium, chromium, molybdenum, tungsten, vanadium, and their compounds are quite common components in the powder metallurgy industry.

In addition to dust, the workers of the mixing departments are exposed to medium-frequency noise occurring during operation of milling and mixing equipment and reaching 80-85 dB.

Considerable dust formation is observed in the pressing department when ejecting the articles from the press dies, loading the hopper of the presses with the charge, and manual proportioning of the charge.

The dust content in the working zone of the press operated during the manufacture of powder metallurgy articles averaged about 15 mg/m^3 and was slightly lower than in the working zones of the mixing department. A high dust content was noted also in the working zone of the pressing department during manual proportioning of the charge by the volume or weight method. Thus, during manual proportioning of an iron–graphite charge the dust content reached 19.5 mg/m^3 and when proportioning the charge for ferrobarium magnets 26.0 mg/m^3.

Considerable dust formation is observed during manual loading of the hopper of the presses: the dust content when performing this operation was 89.3-58.9 mg/m^3 on the average. Sizing of the articles is accompanied by less dust formation, the average dust concentration being 1.7 mg/m^3.

It should be noted that the level of dust production during pressing of metal powder articles is affected by the character of the equipment used and the amount of plasticizer in the charge, and also its viscosity. When pressing large articles and those produced from coarse powders large amounts of plasticizers are usually added, and this markedly reduces dust formation. It has also been noticed that pressing of small articles from a charge with a low content of readily evaporating plasticizer is attended by greater dust formation. These factors apparently explain the low dust content of the air in pressing of large articles on Orenburg hydraulic presses and the comparatively high dust content when mechanical presses are used.

In the case of a high content of plasticizers (alcohol, gasoline, rubber dissolved in benzine) in the charge, their vapors can be released into the air of the working zone. An increase of the amount of organic plasticizers in the charge promotes an increase of the content of hydrocarbons in the working zone during heat treatment of the articles.

The pressing department is characterized by high-frequency noise generated during pressing and sizing of articles. The intensity of the noise depends on the design of the press and varies within 79-100 dB. A weaker noise is noted during operation of hydraulic presses and a stronger noise during pressing on mechanical presses. The noise level also increases with an increase of the power of the press. It has been established that pressing of articles on Orenburg hydraulic presses is accompanied by a noise level of 80-85 dB and on mechanical presses of type RM, 79 dB; PV-474, 80 dB; TM and K-032, 79 dB; P-32, 89 dB; P-16, 99 dB; and KR-1, 100 dB.

The intensity of the noise in the pressing department in shop areas at a considerable distance from the presses reaches 85 dB.

It should be pointed out that the pressing and sizing of powder metallurgy articles place a considerable strain on the operators, since they must pay close attention to the rhythm of operation of the press.

A high dust level is noted during operations in the heat-treatment department. The dust content, reaching tens and sometimes hundreds of milligrams per cubic meter of air, stems from loading powdered materials and articles into boats and crucibles for heat treatment, and unloading them, on open tables not equipped with a cover or exhaust ventilation. The dust content decreases substantially when these operations are performed in areas equipped with such devices. Tens and hundreds of miligrams of dust per cubic meter of air were found during unloading of articles from the boats after sintering in electric furnaces and during separation from the packing material.

During high-temperature treatment of articles and materials containing elements with low melting points, the evolution of a small amount of highly dispersed condensation aerosols, which settle on surrounding objects, is observed in the working zone of the operators near the ports of the furnace. Their formation is due to partial vaporization of metals under the effect of the high temperature in the furnace and subsequent condensation of the metal vapors after entering the zone of lower temperatures.

In addition to dust, the workers of the heat-treatment department can be exposed to a high air temperature and infrared radiation. During sintering of articles and heat treatment of metals at 800-3500°C in electric furnaces of the Tammann, Globar, TsÉP-220, TsÉP-214, TsÉP-356 types, and others in a protective atmosphere of hydrogen, argon, helium, and nitrogen, and also in tunnel- or compartment-type muffle furnaces (OKB-194-1, OKB-197A, G-30, K-130), the air temperature in the working zone of the furnace operators during the fall and winter reached 23-30°C and during the summer 35-40°C. The level of infrared radiation near the charging port of the Tammann electric furnace reached 0.5-1.5 cal/cm^2 per minute and near the open ports of the compartment- and tunnel-type electric furnaces increased to 7-15 and sometimes to 20 cal/cm^2 per minute.

An important role in the formation of an unfavorable microclimate in the heat-treatment shop is played by thermal radiation from the surfaces of the electric furnaces, whose temperature as a consequence of poor insulation increased to 40-220°C. The TsÉP-356 and TsÉP-220 prototype industrial furnaces are the most unfavorable in this respect. During the measurement of the temperature and loading and unloading of articles and materials, the working space of the furnace, too, is a source of thermal radiation.

Thermal radiation from the furnaces extends to other structures (floor, walls, hoods over the charging ports of the furnaces), which heat up to 30-40°C and serve as secondary heat emission sources. Workers of the heat-treatment department are also exposed to temperature differences, which occur in connection with different intensities of heat released and the presence of open apertures in the buildings (doors, windows).

The effect of dust on the workers can be aggravated by the harmful action of carbon monoxide, nitrogen oxides, sulfurous gases, and toxic substances, the content of which near the ports of the furnaces and during oil impregnation, sulfurization, and quenching of articles can exceed the maximum allowable concentrations.

In the workshop dust is the principal untoward factor of the industrial environment. It is formed as a consequence of the grinding of finished articles and rough grinding on abrasive wheels, and also during machining on lathes and milling machines. In this connection, particles of the metals and alloys handled predominate in the dust of mechanical departments. The dust content in the breathing zone of workers during dry grinding increased on the average to 96.8-146.4 mg/m^3 and dropped considerably with wet grinding. In these cases, and also when machining articles on lathes and milling machines, the dust concentration only slightly exceeded the allowable norm.

Dust formation in the workshop is as a rule accompanied by medium-frequency noise, not exceeding, however, the level permitted by sanitary laws.

Other powder metallurgy articles are also produced by the method described (for example, ferrites, hard-metal articles, refractory compounds). Dust, whose composition is determined by the type and purpose of the sintered articles, comprises the main deleterious factor at all production stages.

In the manufacture of ferrobarium and small magnesium–manganese ferrites containing iron, manganese, magnesium, and barium, the dust concentration in the working zones where charges were prepared and their components weighed out reached 2.5-24 mg/m^3. The pressing of articles resulted in dust contents of 5.3-48.0 mg/m^3 (average 20.6).

Approximately the same working conditions were observed by the authors of [132, 283] during the production of various grades of ferrites prepared from iron, manganese, chromium, magnesium, aluminum, zinc, nickel, and vanadium.

A high air dust content was noted during mixing and screening of ferrite powder and oxides. The dust content during these operations increased to 32 mg/m^3; the content of manganse oxides in the dust exceeded the maximum allowable concentration by a factor of 6 and of chromium oxides by a factor of 4-30.

A high dust concentration was recorded in the pressing of ferrites, during operation of the press and during ejection of parts involving handling the die. The maximum content of manganese oxides in these cases exceeded the maximum allowable concentration (MAC) by a factor of 4.5 and the average more than twofold.

When preparing charges and pressing articles containing nickel its concentration in the air reached 0.4-20 mg/m^3.

Dust is produced also during machining of ferrites on circular, face grinding and drilling machines, and air turbines. Maximum dust production was found during dry grinding of ferrites with an air turbine.

During chamfering of ferrite rings the dust level reached 8 mg/m^3, and the manganese content was almost double the allowable concentration. A lower degree of dust formation was noted during wet grinding of ferrites on circular and face grinding machines, but even in this case the amounts of the most toxic components – chromium and manganese – exceeded the maximum allowable levels. The dust produced during the manufacture of ferrites is very fine; the quantity of dust particles measuring not more than 2 μ in most processes reached 85.5-92%.

The release of excess heat, and toxic gases and vapors is characteristic of processes of heat treatment of materials and articles.

The unsatisfactory condition of the ambient atmosphere in the production of ferrites is linked with the factors noted earlier. These are primarily the imperfection of the manufacturing process, lack of mechanization of fabrication operations, inadequate sealing of the dust-producing equipment, and inefficiency or absence of local exhaust ventilation.

According to literature data [283], the working conditions in the production of nickel–zinc, manganese–zinc, and barium ferrites are also characterized by higher dust formation at the stage of mixing of the ferrite components. In the mixing department the air dust content during manual loading and unloading of the materials from the grinding equipment and their screening without covers or local exhaust ventilation reached high concentrations. High dust formation was observed also during pressing and sintering of the articles. In addition to dust, the workers of the heat-treatment department were exposed to sulfur dioxide, arsenic pentoxide, and arsenic trioxide in amounts of up to 3.5 ml/liter of air. These gaseous products were released into the room of the heat-treatment area through the loading and unloading ports of the furnaces, and through the openings for the Silit rods. Sulfur dioxide reacting with atmospheric moisture was the source of high contents of sulfuric acid (up to 8.0 mg/liter) in the shop.

During heat treatment of finished products a high content of carbon monoxide and hydrocarbons, formed as a result of incomplete combustion of organic substances (paraffin wax, polyvinyl alcohol) added to the charge as plasticizers, was recorded in the air.

Sintering of the articles was attended by a rise in air temperature owing to the emission of heat by the furnaces and the further release of heat by the articles as they cooled in the shop.

In the production of hard metals and hard metal articles the manufacturing process includes the production of tungsten and titanium carbides, mixing of these powders with powdered cobalt, and pressing, sintering, inspecting, and packing of articles. In some instances partial machining of the articles is stipulated.

Relatively favorable working conditions were noted in those shops manufacturing sintered hard metal articles where the main operations were partially mechanized and performed with the equipment sealed and with exhaust ventilation at places where dust was liberated. According to our determinations, in the milling and mixing department during the loading of the boats, milling, weighing, and other operations, the dust content was within 2-13 mg/m^3.

During manual proportioning of the charge and pressing of articles on semiautomatic presses the dust content was 3-5 mg/m^3. During loading of the articles into boats and their separation from the packing material after heat treatment the dust content was 1-5 mg/m^3; during inspection of the articles and their cleaning with emery paper the content was 1-9 mg/m^3, and in the case of mechanical grinding and other types of cleaning, 3-11 mg/m^3. Since in the manufacture of hard-metal articles a solution of rubber in benzine is added to the charge as a plasticizer, the hydrocarbon content of the air was determined in certain working zones. It was found that the concentration of these substances in the air of the slurry department, drying shop, and benzine department (where the solution of rubber in benzine is prepared) was within 0.1-0.36 mg/liter, i.e., almost at the maximum allowable level. The carbon monoxide level in the working zone of the furnace operators during sintering of articles also did not exceed the MAC.

The working conditions associated with the effect of noise and a deleterious microclimate in the industry did not differ under investigation from those described earlier.

Similar data on working conditions were obtained at another hard-metal plant, where the main operations were also mechanized. [315]. The average air dust content in this case was within 3-47.7 mg/m^3. Only in the milling and mixing operations did the dust concentration increase to 130 mg/m^3.

The average tungsten content during the main operations ranged from 1.4 to 3.5 mg/m^3 and the content during carbonization and milling, 1.8-2.0 mg/m^3. Sealing off the screens made it possible to reduce the air dust content to about a quarter, and the use of automatic pressing reduced dust pollution by half (from 15.9 to 7.4 mg/m^3).

According to [114,124, 199, 217, 254], dust pollution of the air in the mixing department reached 42.2-83.0 mg/m^3 and in some cases even higher. Less dust formation (of the order of 5.8-18.9 mg/m^3) was noted during pressing. Benzine vapor was detected in the air of the working zone during mixing of the components with rubber and pressing of articles.

Sintering of the articles was accompanied by an increase of air temperature, which in the working zone of the furnace operators reached 32-40°C. The evolution of large amounts of carbon monoxide was observed in this zone during sintering. Hard metal and abrasive dust (up to 9-16 mg/m^3) was released into the air of the working zone during machining of the articles. Similar data are given in [93].

On the basis of what was presented in this chapter we can conclude that at all stages of production of sintered articles the main untoward factor is the mixed dust of the starting materials, intermediate products, and also the finished articles. The level of dust production depends on the degree of mechanization and automation of production operations, dustproofing equipment, and ventilation of the sites of dust release. A high air dust content is associated usually with milling and mixing, loading and unloading of articles and materials from boats, and their heat treatment.

Considerably less dust pollution of the air is noted during operations in the pressing and machining of the articles. During pressing the air dust content decreases considerably when the charge is moistened with plasticizers and also when the articles are moistened before grinding.

Apart from dust, workers engaged in powder metallurgy production can be exposed to the effect of medium- and high-frequency noise from the equipment, high air temperature, infrared radiation, and gaseous substances. The state of the microclimate in the production rooms of the heat-treatment department depends to a considerable extent on the effectiveness of insulation of the outside surfaces of the furnaces, construction of the charging ports, and degree of cooling of the articles being unloaded.

A high air dust content is recorded not only in the immediate vicinity of the operating equipment but also at neutral points in the production rooms. In this case the sources of dust formation can be particles of materials that settled out from the aerosol onto the floor and structural members of the building and equipment, and particles introduced into the shop atmosphere by air currents. Of essential importance in this case is the transport of powdered materials, which is sometimes accomplished manually without appropriate precautionary measures.

During manual transport of powdered metals and articles the workers of many departments experience considerable physical strain which can promote the inhalation of high doses of dust.

With respect to the nature of the deleterious substances being released and other adverse factors, the working conditions in various shops of the powder metallurgy industry are similar in many respects to the working conditions in the production of powders of metals and their compounds (refractory alloys, hard metals, iron powders), which can be explained by the similarity of many production operations related mainly with crushing, milling, screening, and heat treatment of various materials. Operations connected with the mixing of powder with plasticizers, pressing of articles, heat treatment, machining, and other operations are absent in the production of powdered materials. At the same time, in the manufacture of powdered materials operations involved in the classification of the finished powdered substances constitute a larger proportion of the production cycle. The fineness of the air-suspended dust and its concentration in the air of the working zone are directly related to this.

In the manufacture of the various products described here the level of dust pollution is generally similar. This is obviously due to the use, in many cases, of the same or similar batch-type small-capacity equipment, which is often operated by hand. The similarity of dust concentrations in the air of working zones is also linked with the level of mechanization and automation of operations, which in most cases is still low.

The chemical composition of the dust produced during the manufacture of metal powders is distinguished by greater stability than in the manufacture of sintered parts. This dust usually has fewer components. At the first stages of the production process the starting materials are released into the air and at the final stages, the dust of the finished product.

The shape of the dust particles suspended in the air duplicates to a considerable extent the shape of the particles of the powdered materials [95]. The dust at powder metallurgy plants in most cases contains solid particles having the shape of fragments with polymorphous sharp edges.

In the production of certain powder metallurgy parts the technology calls for a certain shape and size of the particles. Powdered materials with spherical particles are used in the manufacture of sintered filters. Their size determines the porosity of the filters produced. Powders of carbonyl iron and of other metals produced by decomposition of carbonyls, having a unique onionskin structure and a spherical surface, can be used for special purposes. It has been noted that a decrease in particle size of powder metallurgy materials has a favorable effect on their strength and other properties. In this connection, in the production of sintered articles the material is as a rule milled in various types of equipment. A decrease in particle size in the manufacture of powder metallurgy materials can in turn be accompanied by increased dust formation.

The shape and size of particles of powdered materials are determined largely by the method of their production. Metal powders are manufactured primarily by mechanical disintegration of metals (filling, crushing, milling, eddy-mill comminution), condensation of metal vapors, chemical or electrochemical deposition (electrolysis of aqueous solutions or salts), spraying or atomization of liquid metal with steam, compressed air or other gas, and dissociation of metal carbonyls.

The production of metal shot by pouring molten metal onto screens has become widespread in industry.

The manufacture of dry multisized powders by mechanical disintegration of metal is used most widely. Many types of metal powders are produced by this method. The production operations involved in their manufacture by this method do not differ substantially from those performed in the preparation of powders in the powder metallurgy industry. This process is generally attended by considerable dust formation, reaching tens and hundreds of milligrams of dust per cubic meter of air [95].

Comminution of metal powders results in the formation of dust-particles of different shapes. In the case of milling of malleable materials in ball mills flat particles are formed. Powdered materials obtained by electrolysis have particles of dendritic structure and complex shape.

Rounded particles are formed in the atomization of metals with a high surface tension. In the case where molten materials with a lower surface tension are atomized, the particles acquire the shape of fragments with sharp edges. The shape and size of particles obtained under identical conditions of mechanical comminution depend considerably on the electronic structure and crystal structure of the substance.

Thus, the conditions of formation of the dust of metals and their compounds at powder metallurgy plants are closely related with the comminution of the materials and their transport. However, in certain thermal operations the liberation of condensation aerosols into the area of the working zone of the furnace operators is possible. This occurs during heat treatment of metal powders under high-temperature conditions (preparations of compounds and alloys). Also possible is partial vaporatization of metallic materials with their subsequent condensation after leaving the charging ports of the furnace and entering the zone of lower temperatures. As a result, small amounts of a highly dispersed condensation aerosol are formed, whose composition corresponds to that of the starting materials, but with a predominance of elements with lower melting points.

CHAPTER III

Morbidity and State of Health of Workers Exposed to Metallic Dust

Metallic dust as an etiologic factor in the development of various pathological states of the organism has attracted the attention of investigators for a long time. At present considerable data have been accumulated on the untoward effect of various types of metallic dust on the health of workers at plants of ferrous and nonferrous metallurgy, machine building, and in arc welding operations.

The adverse effect of metallic aerosols increases the morbidity of workers with temporary disability and is the cause of a number of occupational diseases affecting internal organs, primarily respiratory organs.

At industrial plants the effect on workers of dust having a mixed composition (including aerosols of pure metals, their oxides and compounds), is well noted; an effect from the polymetallic dust is also possible. However, the problem of the unfavorable effect of individual components of the aerosol, particularly changes in workers induced by pure metals or their oxides and compounds, is poorly elucidated in the literature. Difficulties sometimes arise in determining the cause of the pathological state and the role of metallic dust in the development of diseases because of concomitant effects of other factors in the industrial environment (noise, unfavorable microclimatic conditions).

Morbidity with Temporary Disability among Workers Exposed to Metallic Aerosols

An analysis of morbidity with temporary disability among workers in powder metallurgy shops indicates a higher level of morbidity in these shops in comparison with that of the general plant, and in comparison with the morbidity in auxiliary shops where the effect of dust on the workers is eliminated. Influenza, angina, diseases of the upper respiratory tract, bronchitis, acute gastrointestinal disorders, and diseases of the heart and nervous system are the diseases which occur most frequently.

In some shops there is a high incidence of diseases of the female genitalia and cases of ulcers. The highest morbidity is found among mixers and millers exposed during work mainly to the effect of metallic dust and also among batchers, pressers, and grinders during whose work the effects of metallic dust and noise predominate. A slightly less morbidity was observed among the sizing-press operators, furnace operators, and repairmen who were exposed less to dust and noise.

Diseases of the respiratory organs (upper respiratory diseases, bronchitis, pneumonia) predominated among workers of the first two groups. In addition to these diseases, in the pressing department there was a high incidence of angina, rheumatism, and diseases of the heart and gastrointestinal tract, including ulcers.

In the group of sizing-press operators, whose work is characterized by considerable neuromuscular tension in addition to the effect of dust, there were numerous diseases of the peripheral nervous system.

Cases of influenza, upper respiratory diseases, and diseases of the peripheral nervous system were frequent in the group of furnace operators and repairmen.

Pneumoconioses and dermatitides were recorded in addition to the indicated general forms of diseases.

An increase of the level of morbidity with temporary disability in comparison with that of the general plant was noted also in workers of the iron-powder production shops, who are exposed mainly to the dust of reduced iron and its oxides [106].

According to [215], the level of morbidity was high among machine workers (lathe operators, milling machine operators, turret lathe operators, drillers). No significant differences were noted, either in the level or the form of morbidity, when compared to the morbidity of welders. The frequency of individual diseases (angina, pyrexia, pharyngitis, tonsillitis, tuberculosis, dermatitis) was approximately the same.

Most frequent among welders were bronchitis, conjunctivitis, and gastrointestinal disorders. Classifiable forms of the diseases indicated above were noted frequently at ferrous metallurgy plants as a result of exposure to iron oxide dust [12].

Morbidity higher than in ferrous metallurgy is characteristic of copper melting plants, where copper dust numbers among the unfavorable etiological factors [176]. According to [154], influenza, bronchitis, and acute gastritis were higher among persons exposed to high concentrations of copper dust than in other shops.

Predominant at lead metallurgy plants are lead poisoning and diseases of the digestive organs caused by prolonged exposure of the workers to lead-containing dusts and lead vapors [143]. The proportion of respiratory diseases at such plants is at the same level as at other metallurgical plants. The incidence of respiratory diseases is higher in the ore shop with a high airborne dust level.

A considerable effect of metallic dust resulting in an increase of morbidity with temperary disability is noted also in other industries [67, 280]. An increase of general morbidity and an increase of the number of respiratory diseases is noted in particular in workers employed in the vanadium industry [280].

A special feature of morbidity with temporary disability in workers exposed to cadmium oxide is that liver diseases rank high in number of cases and days of disability [62]. Among these diseases are hepatitis and hepatocholecystitis, occurring three years after starting work in contact with cadmium.

These data on morbidity with temporary disability among workers subjected mainly to the effect of metallic dust permit the conclusion that metallic dust can be an etiologic factor adversely affecting the workers' health. A high incidence of influenza, upper respiratory diseases, bronchitis, and sometimes pneumonia is characteristic of powder metallurgy production, and also of other plants with a high content of metallic aerosols. Pneumoconioses and a rather high level of pathology of digestive organs and purulent lesions of the skin are found during prolonged work under conditions of high dust pollution.

It is clear that the pattern of diseases, and the increase in the number of nonspecific diseases such as influenza, angina, and catarrhal conditions of the upper respiratory tract can probably be attributed to a decrease in the immunobiological reactivity of the organism. This conjecture is based on the results of investigations [240] which established that industrial toxins can lower the barrier function of the organism considerably. It should be noted that the number of nonspecific diseases and chronic occupational poisonings can increase under the action of metallic aerosols having a pronounced toxic effect (lead, mercury, manganese).

State of Health of Workers Exposed to Metallic Dust

Metals can be divided into three groups depending on the degree of affection of various organs and course of the pathological process (Fig. 11).

The first group includes metallic aerosols causing pathological phenomena, primarily of the respiratory organs (iron, tungsten, niobium, tin, etc.).

The second group includes aerosols of metals and their compounds affecting respiratory organs and causing dysfunction of internal organs (chromium, molybdenum, vanadium, nickel, cobalt, copper, silver, cadmium, etc.).

Period	IA	IIA	IB	IIB	IIIB	IVB	VB	VIB	VIIB	VIII B			IIIA	IVA	VA	VIA	VIIA	VIIIA
1	H				1	2	3	4	5	6								He
2	Li	Be											B	C	N	O	F	Ne
3	Na	Mg											Al	Si	P	S	Cl	Ar
4	K	Ca	Cu	Zn	Sc	Ti	V	Cr	Mn	Fe	Co	Ni	Ga	Ge	As	Se	Br	Kr
5	Rb	Sr	Ag	Cd	Y	Zr	Nb	Mo	Tc	Ru	Rh	Pd	In	Sn	Sb	Te	I	Xe
6	Cs	Ba	Au	Hg	La	Hf	Ta	W	Re	Os	Ir	Pt	Tl	Pb	Bi	Po	At	Rn
7	Fr	Ra			Ac													

Fig. 11. Change of the toxic properties of elements as a function of their electronic structure: 1) s elements; 2) sd transition metals; 3) sp elements; 4) acute poisons; 5) chronic poisons; 6) pneumoconiotic effect.

The third group is made up of metals having a toxic effect which can cause acute and chronic poisoning (for example, mercury, lead, manganese).

The most frequent occupational disease occuring in workers during prolonged inhalation of the dust of certain metals is pneumoconiosis – dust disease of the lungs, underlying which are sclerotic and other changes effected by the deposition of various types of dust.

As long ago as the 1930's mention was made in the handbook of Henke and Lubarsch [469] of siderosis, caused by the inhalation of iron dust, as an independent form of pneumoconiosis.

In an examination of the state of health of workers of powder metallurgy plants by the clinical department of the Kiev Institute of Industrial Hygiene and Occupational Diseases [145], chronic bronchitis was noted in 17.5% of the cases in persons exposed primarily to the effect of the dust of iron, aluminum, beryllium, i.e., metals having a fibrogenic action. Peribronchitis accompanying chronic bronchitis and displaying roentgenologically an intensification of the pulmonary picture was found in 2.9% of the cases, mainly due to the effect of fibrogenic dust. In this same group 7.5% of those examined had pneumoconiosis, which developed in persons working in powder metallurgy for 5-12 years.

Changes in the air passages, in the forms of rhinitis and rhinopharyngitis, which in some cases were attended by a decrease of olfaction, were frequently noted among these contingents of workers. A decrease of olfaction was observed also in workers of iron powder shops who had been exposed mainly to the dusts of iron oxides and iron powders [106, 145].

Batsheva and Miller [15] revealed pneumoconiotic changes in the lungs of workers who had been inhaling arc welding dust. Later, other authors [63, 87, 133, 142] also indicated the possibility of changes in the lungs under the action of the dust of iron-containing arc welding aerosols. Clinical observations were confirmed by data of pathologo-anatomic investigations [90]. Siderosis, which in the presence of roentgenomorphological changes had occurred without symptoms, was found in steelworkers with more than 17 years' service (in 9 out of 17 workers). Similar data were obtained on examining workers engaged in the production of ocher, which contained 75-90% ferric oxide and 2-5% SiO_2 [215].

Sometimes the discrepancy between the roentgenological picture and general state of the workers permits attributing the changes revealed on the roentgenograms to the x-ray contrast of iron dust.

On examining 130 workers employed in the production of refractory compounds who were exposed to the dust of chromium, zirconium, and molybdenum oxides and of chromium carbide, zirconium carbide and boride, molybdenym disilicide, boron and silicon nitrides, and boron and silicon carbides, we detected in some of them a decrease of olfaction. We did not observe in these workers a difference in the olfaction threshold level to cause a substantial reaction in the olfactory and trigeminal parts of the olfactory analyzer. However, the majority of workers even with short service periods (up to 5 years) had a diminshed olfaction threshold. Adaptation to threshold stimulation in them occurred in the majority of cases (54.7-72.5%) after 3 min, i.e., with some delay.

Recovery of the sensation threshold lagged especially in workers exposed to silicon carbide dust, after stimulation equal to five times the threshold level. In 52.2-59.0% of the examinees the recovery time of the sensation threshold was almost double that of workers exposed to metallic dust and boron carbide dust.

Similar data were obtained by other authors in medical examinations of hard-metal production workers [53, 437, 438, 500, etc.]. An examination of the state of health, at these plants, of workers who had been exposed to the dust of tungsten, titanium, cobalt, and titanium and tungsten carbides revealed changes on the part of the air passages. Irritations, rhinitis, and rhinopharyngitis, both atrophic and subatrophic, were observed. In some, asthmoid bronchitis occurred, and pneumosclerosis had developed.

The most detailed description of the state of health of workers at hard-metal plants is given by G. I. Bunimovich and N. N. Buravleva [53], who observed 18 workers in a clinic for examination and treatment, and also examined another 185 workers who had been in contact with the dust of tungsten, cobalt, and titanium compounds in concentrations exceeding the allowable.

The clinical manifestations of pulmonary diseases of the 18 patients fell into three classifiable forms: asthmoid bronchitis (8 persons), pneumonia (5 persons), and chronic interstitial pneumonia (5 persons). The patients complained of a paroxysmal-like cough with expectoration and asthma accompanied by difficult exhalation. The disease developed gradually and began with a cough with difficultly expectorated sputum. Later the cough acquired an asthmoid character, and during the last year before hospitalization attacks of asthma occurred. Slight expansion of the chest, decrease of the limits of the lungs, ascultatively slow exhalation, and disseminated dry rales were noted; percussively the tone did not change. Expressed pathology was not observed on the part of the cardiovascular system; the arterial pressure was normal. Moderate eosinophilia, up to 9%, was noted in the blood.

Slight intensification and deformation of the pulmonary picture was observed in a roentgenological investigation of the patients. The diaphragm was slightly constrained, the sinuses were free. The attacks of coughing and asthma that occurred in the patients were controlled by ephedrine and epinephrine injections. The asthma attacks ceased in seven patients after they were no longer in contact with the industrial dust. Only in one, who continued to work, did the asthma attacks continue to recur and phenomena of emphysema and pulmonary insufficiency occurred.

In five workers of powder metallurgy shops the aforementioned authors observed the development of dust lung. The roentgenological data resembled pneumoconiosis quite closely. The roentgenograms showed fibrosis with numerous small, primarily pointlike, nodular shadows, most densely arranged in the middle and lower pulmonary lobes. In addition to this, numerous ringlike shadows of indurated bronchi were noted. The roots of the lungs were also indurated and had petrifacts. A functional investigation of respiration did not uncover appreciable disturbances. The vital capacity of the lungs was 2600 cm^2 at a norm of 3100 cm^2.

Dynamic observations for three years showed stability of the changes noted on the early roentgenograms, without explicit progress.

Many investigators have noted pneumoconiotic changes under the effect of the refractory compounds under consideration [201, 202, 433, 487, 491, 494]. Characteristic in this case was the presence of granulomatous and conglomerative changes in the lungs of 2% of the total number of examined workers engaged in the production of hard metals, although some of them had worked earlier under dusty conditions.

Phenomena of hypoacidic gastritis predominated in persons in contact with cobalt chloride, whose adverse effect on the gastric function was confirmed by the investigations of Kaplun [119, 120].

Bushueva's examination [53] of 52 workers revealed in many of them a large number of cases of subatrophic upper respiratory diseases and chronic inflammations of the paranasal sinuses; impairment of olfaction was noted exclusively in persons with diseases of the paranasal sinuses.

Thus three types of pulmonary diseases were found among workers of hard-metal plants: asthmoid bronchitis, pneumoconiosis, and interstitial pneumonia. They developed primarily in workers with long service who had been subjected to the effect of high concentrations of mixed dusts of cobalt and tungsten and titanium carbides.

Kaplun and Mezentseva [122] analyzed data yielded by periodic medical examinations of 283 workers of hard-metal plants working under conditions of constant dust liberation, in some of whom incipient phenomena of diffuse pneumosclerosis were established roentgenologically. The results of the investigation of a second group of workers (247 persons), who had been in contact with cobalt dust, indicated that the main complaints were loss of appetite, nausea, coughing, and diminution of the sense of smell. A medical examination revealed in 117 persons lesions of the upper respiratory tract (hyposmia, and in some cases anosmia), in 35 persons chronic bronchitis, and in 33 persons phenomena of incipient pneumosclerosis. Many suffered from hypotension and changes of the blood composition.

Epistaxis, hyposmia, chronic coryza, and other phenomena occurred in workers producing ferrites [283]. During heat treatment of charge materials containing barium nitrate phenomena of acute poisoning of the lungs by nitrogen oxides were noted in workers in this industry.

In the literature there are indications of the possibility of the development of pneumosclerosis and other toxic phenomena in workers engaged in the production of certain metal powders. An examination of workers in a shop producing metallic molybdenum and molybdenum wire who had been exposed to molybdenum trioxide for 4-7 years roentgenologically revealed pneumoconiosis in 3 of the 19 examined [224]. The workers complained of dyspnea, general weakness, fatigue, and some had a moist cough in the mornings. An objective investigation revealed disseminated dry rales in both lungs and pulmonary murmur with a bandbox sound [224].

Pneumoconiosis occurred in workers involved in the production of tantalum by the reduction of potassium fluotantalate with metallic sodium [97]. In three out of 12 workers exposed to a highly dispersed tantalum aerosol in concentrations reaching tens and hundreds of milligrams per cubic meter of air, initial phenomena of pneumosclerosis were found which were similar to those occurring after exposure to titanium, tungsten, and iron.

Aerosols of metals of the second group are characterized by a slightly different effect. An examination of the state of health of workers at powder metallurgy plants exposed to the effect of aerosols of nonferrous metals (copper, cadmium, silver, nickel, cobalt, chromium) and their oxides revealed, as in the case of metals of the first group, chronic bronchitis [144, 145]. Phenomena indicating an irritating effect of these dusts on the bronchi were noted in 10.9% of the examinees. In addition, disturbances in the activity of many systems and organs were found in this contingent of workers.

An examination of the cardiovascular system of the workers revealed a predominance of arterial hypotension (18.5%); hypertension was found only in 9.6% of the examinees. Diseases of the digestive organs were slightly less than in mass medical examinations of other occupational groups, with the exception of the higher frequency of moderate disorders of the liver. A tendency toward lymphopenia was also noted in these workers. The number of leukocytes did not exceed 4500-4700/mm^3 of blood in 12.7% of those examined.

Changes in the nervous system in the form of an asthenovegetative syndrome, neurocirculatory dystonia of the hypotonic type, and, more rarely, vegetative-endocrine dystonia were found in many of the examinees. The development of these pathological conditions in some cases was definitely affected not only by dust but also by such additional factors as high-frequency noise, high temperature, infrared radiation, and gaseous toxic substances (carbon monoxide, etc.). However, in a number of cases these factors were absent, and therefore the authors are inclined to attribute nervous disorders to the effect of ferrous and nonferrous metal dust. Phenomena of neuromyosites as a consequence of overstrain of the neuromuscular apparatus of the arms were noted in 9% of the examiness.

Diseases of the upper respiratory tract (nose, nasophranyx, and larynx) were established in a considerable number of those examined. Hypertrophic forms of rhinitis and pharyngitis were observed in workers with a short service time, whereas with an increase of service time the frequency of subatrophic forms of diseases of the nose, nasopharynx, and larynx increased.

Among persons exposed to rare and nonferrous metals, a decrease of olfactory sensitivity, primarily in the form of hyposmia of the first degree and, more rarely, moderate hyposmia of the second degree, was noted in 12.5% of the examinees. The toxic effect of nonferrous and rare metals was accompanied by diminution of the auditory function. Pneumoconiotic and general toxic changes in workers under the action of the dust of rare and nonferrous metals were observed also by other authors.

Aerosols of vanadium pentoxide and other compounds caused acute and chronic poisoning [89, 170, 191, 276, 427, 428, 454, 456, 517]. In the case of acute poisoning by vanadium pentoxide marked catarrhal inflammation of the mucous membrane of the air passages (nasopharyngitis, pharyngitis, bronchitis) and of the mucous membrane of the eyes was observed. An admixture of blood was noted in the sputum and nasal discharges as a consequence of injury to the vascular wall. Expiratory dyspnea, retrosternal pains, neurotic states, and skin rash were characteristic. Frequently there was a coating on the oral mucosa as a consequence of salivary excretion of vanadium. The metal was also excreted in the urine.

An examination of persons exposed to vanadium pentoxide and trioxide in small concentrations (2.0 mg/m^3) demonstrated hypertrophic, substrophic, and atrophic rhinopharyngitis, pronounced hyperemia of the mucous membrane of the nose and pharynx, and cases of chronic bronchitis in workers with six and more years of service.

In addition to these symptoms, affections of the central nervous system (tremor of the fingers and hands) and accent of the second tone of the pulmonary artery as a consequence of dysfunction of the respiratory organs (bronchitis, emphysema, etc.) were found in pronounced chronic poisoning. The latter was also caused by the chronic effect of the dust of ferrovanadium.

Affections of the respiratory organs, particularly of the upper respiratory tract, accompanied in many cases by complete absence of olfaction, and also bronchitis and pneumosclerosis were observed in workers exposed to cadmium oxide. Among the complaints of the workers examined we need point out disturbance of a dyspeptic nature, and an objective investigation revealed in some of them chronic diseases of the liver accompanied by its enlargement and painfulness on palpation. A tendency toward lymphopenia was observed in many of the workers – the number of leukocytes was less than 4500. Symptoms of affection of the central nervous system were also pronounced [62, 409, 465].

A toxic effect is caused by excessive amounts of copper and its compounds which, when taken up by the organism, produce strong irritation of the mucosa of the air passages and gastrointentinal tract [153]. The untoward effect of copper salts, cupric oxide, and pure copper on workers in whom phenomena of "metal fume fever" occurred, has been described [4, 54, 103, 145, 146, 153, 154, 377, 421].

The most toxic of the chromium compounds are the hexavalent and trivalent. As Chekunov indicates [161], divalent compounds and chromium are not particularly toxic. Chromic acid, chromates, and dichromates irritate and burn the mucous membranes and skin, causing ulcerations; cases of ulceration and perforation of the nasal septum are common following inhalation of aerosols of these compounds. The compounds in question also have a general toxic effect, affecting mainly the gastrointestinal tract. Chromium compounds give rise to hypersensitivity of the organism and cause attacks resembling those of bronchial asthma. A high rate of cancer of the air passages is noted among workers exposed to chromium, which is related with the specific action of chromium [19-21].

A large number of workers exposed to chromium oxides also have inflammation and swelling of the larynx and vocal cords and it has been concluded that the changes are proportional to the concentration of the dust of chromium compounds in the air [161].

In acute intoxication by potassium dichromate dust dizziness, vomiting, chill, increase of pulse rate, and pains in the gastric region were manifested.

Sensitization occurred in the organism when working with chromium oxides. Cases of attacks of bronchial asthma in persons working in contact with chromium are described in [58]. Cases of bronchial asthma in workers of heat-treatment shops exposed to ferrochromium for more than 10 years are indicated.

The symptoms of chronic poisoning by chromium compounds have been described. In this case the condition of the workers was characterized by headaches and emaciation; inflammatory and ulcerative diseases of the gastrointestinal tract and pneumonia occurred. In workers of the chromium industry the urine often contained albumin and blood; cases of pneumosclerosis, changes on the part of the myocardium, and gastritis were not uncommon. Cases of hypochromic anemia and leukocytosis were also described in this group of workers [88].

There is information [393, 396] on the possibility of the development of various forms of cancer under the effect of chromium compounds. The average length of service of workers suffering from lung cancer is 12-22 years. The tumor is most often localized in the right lung; this is usually a squamous-cell carcinoma, more rarely adenocarcinoma. Sometimes perforation of the nasal septum occurs simultaneously in the cancer patients along with the presence of increased amounts of chromium in the blood and lungs.

Chromium compounds cause ulceration of the skin [16, 19, 21, 411]. The authors of these studies revealed in 16.5% of the workers hypersensitivity to chromium, due to which eczema can develop after 6-9 months of work. Chromium is readily absorbed through the mucous membranes and skin. It was established experimentally that, upon entering the body, it is deposited in the liver, kidneys, endocrine glands, hair, nails, and teeth. Inhaled chromium dust is deposited in the lung tissues [88, 161]. Treatment of asthmatic attacks occurring as a consequence of the effect of chromium is the same as for bronchial asthma (atropine, adrenaline, dimedrol, etc.).

Kochetkova [149] describes the autopsy material of a female who died after working seven years in contact with cobalt dust. A case of poisoning after milling cobaltous acetate and inhalation of its dust is described [161]. Irritation of the mucosa of the fauces, bloody vomit, marked hypersensitivity of the abdomen, intestinal colic, high temperature, and weakness in the legs were observed an hour after the end of work. Recovery took four weeks. Sometimes inhalation of cobalt dust resulted in bronchial asthma [88, 161].

Cases of skin lesions in workers exposed to cobalt are described in the literature: acute dermatitis in the form of numerous red papules and nodules, and also edema of the hands and exposed portions of the body; sometimes surficial ulcerations, papules, and nodules.

In the case of poisoning by nickel compounds various pictures of toxicosis develop, and pneumonias and dermatitides occur frequently. Some authors [161, 389] consider the occurrence of dermatitides to be an allergic reaction to nickel. It was conjectured that the inhalation of nickel compounds can cause an allergic reaction also in the respiratory apparatus.

It is reported [161] that in a storage-battery factory albumin was found in the urine of almost all workers with a service of more than eight years; in this case the nickel concentrations in the air varied within 0.01-0.07 mg/liter, but since cadmium vapor was also present in the air it was not conclusively established whether kidney damage was due to nickel compounds.

In England 39 cases of cancer of the lungs and paranasal sinuses were recorded in workers of the nickel industry. These diseases were attributed to the effect of nickel carbonyl [161]. Three cases of lung cancer in workers who inhaled only the dust and fume of metallic nickel and some of its compounds but not nickel carbonyl vapor are described in [499]. The skin diseases "nickel eczema" and "nickel itch" occur in nickel-plating workers and those employed in the electrolytic production of nickel. Lesions are most common following contact with nickel sulfate.

An analysis of the data on pathological conditions arising under the action of aerosols of metals of the second group permits the conclusion that their effect is similar in many respects. A characteristic feature of their untoward effect is that in such cases the pathological changes in the workers are not limited to affection of the respiratory organs: The nervous and cardiovascular systems and digestive organs are also involved, and rather frequent lesions of the liver are observed. An allergic state is characteristic, in connection with which cases of asthmoid bronchitis, dermatitides, etc., are common. As a rule the aerosols under consideration have a pronounced irritating effect, and in some cases cause malignancy.

Metals of the third (tentatively formed) group – lead, mercury, manganese, antimony, and others, which under industrial conditions can affect the organism when present in the form of vapor and dust – cause acute and chronic poisonings attended by affection of many systems and organs. These substances belong to the category of inorganic poisons, and their protracted action leads to serious disturbances in the activity of the central nervous and cardiovascular systems, function of the liver, digestive organs, and kidneys (since they are excreted in considerable amounts in the urine). Under the effect of these highly toxic elements and their compounds changes on the part of the peripheral blood are encountered rather often. The liberation of lead into the air and danger of lead poisoning are possible also in the powder metallurgy industry, particularly in the manufacture of certain types of piezoceramic articles [150, 151].

Lead poisoning in industrial workers is primarily chronic. In certain instances lead poisoning can exist for a time almost without symptoms, In other cases disturbances of the central nervous system occur at early stages of intoxication, a blue line appears on the gingival margin, changes occur in the peripheral blood – reticulocytosis and basophilic stippling of erythrocytes, a high content of porphyrins is found in the urine, and the quantity of lead exceeds 0.05 mg/liter. As a consequence of vascular spasms and anemia a characteristic "lead coloring" – a pale-gray color of the face, appears [195].

The most frequent syndromes of chronic lead poisoning are: anemic, gastrointestinal (lead colic), hepatic, astheno-vegetative, and the syndrome of encephalopolyneuritis [260].

Acute and chronic poisoning is possible when working with metallic mercury and with mercury compounds at mercury mines, mercury plants, in the production of measuring instruments, x-ray tubes, mercury pumps, in the production and use of organomercury fungicides, etc. Substances possessing different physical and chemical properties are characterized by different degrees of toxicity. Metallic mercury is the most active industrial poison. At present acute poisoning by mercury vapors occurs quite rarely, being mainly due to accidents as a consequence of faulty equipment. Of main importance is the chronic form of intoxication in which the severity of the symptoms is determined by the duration of exposure and massiveness of the acting concentrations.

Chronic poisoning usually develops gradually, with few symptoms over a long time. Irritability, general weakness, and loss of appetite are observed in this period. Disorders of the activity of the central nervous system become more pronounced later, even to the extent of toxic encephalopathies. Autonomic deviations are combined with dysfunction of the endocrine glands. Lesion of the mucous membranes of the oral cavity (gingivitis, stomatitis), disturbance of the activity of the digestive organs, functional disorders of the cardiovascular system and thermoregulation, sometimes with persistent subfebricity, and irritation of the kidneys with symptoms of lesion of the renal parenchyma are characteristic. In the blood picture moderate lymphocytosis and, in more severe cases, a decrease of hemoglobin are observed. Multiple ulcerations of the gastric and duodenal mucosa, atrophic changes of the liver, and an increase of urination in the initial stage of intoxication and its decrease and even complete cessation in severe cases can occur [260].

The toxic effect of manganese and its compounds and alloys have been studied by many Soviet and foreign authors [80, 163, 184, 365]. Mainly chronic poisoning that occurred during extraction and processing of manganese ore, in the production of dry-cell batteries, production of ferromanganese, high-grade steel, in electric welding operations, etc., were recorded.

Observations show that in persons with manganese poisoning the nervous system is affected most. The main symptoms of poisoning are disorders related with lesion of the cortex and subcortical portions of the nervous system (primarily extrapyramidal), in the presence of which the syndrome of lesion of the motor analyzer gradually develops. The pronounced form of chronic poisoning is characterized by a unique picture of Parkinsonism. The internal organs are also affected in manganese poisoning. Hepatitis with enlargement of the liver, disturbance of carbohydrate and antitoxic functions, frequent pneumonias, and pneumosclerosis with an increase of cardiac insufficiency are noted.

In Norway the incidence of lobar pneumonia in workers engaged in manganese alloy production is four times higher than that in the rest of the population [363, 364]. Before the Revolution a high mortality and morbidity due to lobar pneumonia was recorded in workers of the Chiatura mines. In manganese poisoning there are also gastrointestinal disorders and frequent phenomena of chronic gastritis with a decrease of acidity of the gastric juice, dermatitides occurring as a chronic inflammation, and changes in the blood picture: an elevated content of hemoglobin and a leftward shift of the neutrophil formula [260]. At the same time, the content of prothrombin, pyruvic acid, sugar, and potassium and calcium in the blood did not change. The oxidative coefficient of the urine, glutathione concentration, and Quick test deviated definitely from the norm, indicating disturbance of oxidation processes in the organism and a decrease of the antitoxic function of the liver [80].

The clinical picture of intoxication by aerosols of metals of the third group is characterized by changes on the part of the functions of the central nervous system, parenchymatous organs (liver, kidneys, heart), and gastrointestinal tract. Lesions of the oral mucosa and skin are frequent. Affections of the lungs in such cases are relegated to the background and are characterized primarily by toxic pneumonias. The development of manganese pneumoconiosis has not been proved conclusively, since the pneumoconiotic lung changes described, which are often encountered in manganese ore miners and processors, were related with inhalation of mixed dust containing silica.

Some organometallic compounds also have a pronounced toxic effect; highly toxic organic compounds of mercury, tin, copper, etc., are known. Of great interest are available data on the toxic effect of metal carbonyls on workers, since these compounds are used widely in the chemical industry, and also in the production of pure metal powders by way of the carbonyl cycle.

Cases of acute poisoning of workers by iron pentacarbonyl and nickel tetracarbonyl have been described [327-329, 381, 514]. Inhalation of nickel carbonyl vapor affects the nervous system and also causes pulmonary edema [328]. The patients have severe headaches, dizziness, nausea, vomiting, and impairment of the coordination of movements (uncertain gait). These symptoms are later followed by dyspnea, chest pain, dry excruciating cough, and cyanosis of the lips and ears, with the skin becoming pale and the forehead cold and sticky from perspiration. The patient is restless, the pulse at first is rapid and then becomes feeble. During the first days of the disease the body temperature is normal, but subsequently it can rise. In mild cases poisoning occurs in the manner of "foundryman's fever," and after a time spent in fresh air the symptoms gradually disappear.

Changes of the morphological composition of the blood was noted in those poisoned: leukocytosis and eosinopenia, sometimes leukopenia with agranulocytosis and reticulocytosis. Pathological changes in the liver were added later. The patients complained of the sensation of pressure in the right hypochondrium, pain upon palpation of the liver, and urobilin and urobilinogen were found in the urine. In milder cases these phenomena gradually abated within 10-14 days. In severe cases cardiac insufficiency increases. Death ensues after 10-14 days with a picture similar to poisoning by suffocating gases. Sometimes death occurred after 2-4 days.

The pathologico-anatomic changes are characterized by edema, plethora, and pneumonia, sometimes by consolidation and proliferation of fibrous tissue in the lungs; plethora and small hemorrhages are found in the gastric and intestinal mucosa. Hemorrhages in the adrenal glands and degenerative changes of the liver cells, epithelium of the renal convoluted tubules, cells of the incretory and excretory parts of the pancreas, and of the adrenal glands are observed.

One author [514], investigating the brains of those who died from nickel carbonyl poisoning, found hemorrhages in the white substance of the cerebral hemispheres. The surface of the brain had a rust color.

Nickel carbonyl has an irritating effect on the air passages and is readily absorbed through the skin. On the basis of a study of the morbidity among workers in England it is considered proved that nickel carbonyl causes cancer of the nasal cavity and lungs when these organs are exposed to it for a long time [218, 422].

Some investigators [328, 484, 486] indicate that nickel can be found in the blood and urine of poisoned workers. Among workers exposed to nickel carbonyl in concentrations of 0.003-0.0045 mg/liter, nickel was found in the urine only in 25% of the cases. A 0.05% nickel carbonyl concentration causes feelings of anxiety and 0.5% is dangerous for health [514]. The mechanism of the toxic action of nickel carbonyl, as suggested in [328, 486], consists in the following: entering the organism, it dissociates with the formation of carbon monoxide and nickel ions. The latter form complexes with proteins, enter the blood, then the brain, liver, kidneys, and heart, effecting pathological changes, and are excreted in the urine. Hemoglobin, combining with the carbon monixide, forms carboxyhemoglobin.

Iron pentacarbonyl is also a highly toxic compound [18, 327, 381]. Upon inhalation or entrance per os or through the skin, it causes acute poisoning, accompanied by pulmonary edema and other phenomena. Cobalt tetracarbonyl is much less toxic than nickel and iron carbonyls [18, 325].

There are no data in the literature on the effect of molybdenum carbonyl. The toxic effect of tungsten carbonyl is indicated in Frolova's report [362]. Exfoliation and hyperfragility of the nails and desquamation of the skin were noted in workers having long contact with tungsten carbonyl.

Thus the few existing literature data indicate the possibility of acute and chronic poisoning by metal carbonyls. In this case the toxicity decreases from nickel carbonyl to iron, cobalt, and tungsten carbonyls.

Beryllium also ranks high among toxic metals. Acute poisoning occurs most often under the action of salts of beryllium and its oxyfluoride, but it is possible also upon inhalation of large quantities of the oxide and metallic beryllium [383, 552, 559, 565, etc.]. Van Ordstrand [581] reported on 170 cases of beryllosis with predominant affection of the respiratory organs and skin, and Laskin et al. [490] on 136 cases of acute poisoning. According to the official data presented at a special conference on beryllosis in Massachusetts in 1958, between 1952 and June 1958 there were 606 cases of poisoning recorded in the USA, of which 303 were chronic.

Poisoning occurred most often in the production of beryllium from ores, manufacture of fluorescent lamps, and melting and machining of beryllium. The clinical aspects of acute and chronic beryllium poisoning are described in [20].

In the case of acute beryllium poisoning the patients develop a severe cough, chest pain, dyspnea, cyanosis, and sometimes a state resembling "foundryman's fever"; chills appear, the temperature rises, and shifts on the part of the blood are noted. After the temperature drops catarrhal changes of the upper respiratory tract are observed – laryngitis, tracheitis, and bronchitis. Pneumonia is sometimes not accompanied by upper respiratory disease. In this case dyspnea and cyanosis occur in the presence of a slight fever, infiltrates are identified in the lungs roentgenologically, and intensification of the pulmonary picture is noted. Histological changes observed in the deceased are characterized by hemorrhages and edema of the lungs, histiocytic reaction, and degenerative changes in the liver.

Chronic poisoning when working with beryllium (berylliosis) develops in a period ranging from several months to ten years and more. The disease can appear also a long time after the end of exposure to beryllium [363]. Berylliosis has been reported in persons who did not work with beryllium but lived in the vicinity of plants, the air near which contained high beryllium concentrations [408, 419, 426].

Chronic berylliosis is marked by loss of appetite, sharp weight loss, general weakness, dry cough, headaches with general cardiac insufficiency accompanied by severe cyanosis, enlargement of the liver, and sometimes clubbing of the ungual phalanges. A considerable content of serous fluid is found in the body cavities. Diffuse reticulation with nodular formations is observed in the roentgenological investigation of the lungs.

The chronic process in the lungs causes circulatory disorder ("cor pulmonale") and congestive sclerotic changes in the liver; the spleen enlarges, and thickening of the periosteum of the ribs and the long bones occurs. A pathologico-anatomic investigation reveals chronic pulmonary granulomatosis, emphysema, sometimes pronounced interstitial sclerosis, a nodular process, thickening of the bronchial walls, and catarrhal or purulent bronchitis.

A granulomatous process developing in the lumen and walls of the alveoli, alveolar ducts, bronchioli, and small bronchi with their obliteration is observed most often microscopically in the lungs in chronic berylliosis. In early stages the granuloma consists of histiocytes with a small number of lymphoid and plasma cells. Sometimes multinuclear giant cells of the foreign-body type whose protoplasm contains crystalline formations are found. Reticular and collagen fibers subsequently appear in the granulomas, the number of cells decreases, and sclerotic nodules are formed, which, merging, can form large conglomerates.

Characteristic granulomas can be found not only in the lungs but also in other organs: in the liver, spleen, lymph nodes, and subcutaneous tissue in the case of introduction of beryllium dust through the skin. Beryllium was found on analyzing the lung tissue and other organs, and fluorescent grains were detected upon ultraviolet irradiation in the lung tissues of the deceased [17, 141, 426, 471, 502]. The dust of beryllium and its salts can cause dermatitides, ulceration of the skin, formation of granulomas, pronounced conjunctivitis, and irritation of other mucous membranes [426].

The effect of alkali metals on the organism is manifested mainly in the irritating action on the mucous membranes of the air passages and eyes, and also on the skin; they can cause ulcerations, dermatitides, and burns. Of these metals the general toxic effect of lithium is most pronounced [244]. The symptoms of poisoning following its inhalation are as follows: general weakness, loss of appetite, thirst and dryness in the mouth, sometimes salivation, nausea, vomiting, diarrhea, tremor of the lips, lower jaw, and hands, dizziness, and impaired vision. In more severe cases epileptoid seizures, and sometimes mental disorders and coma occur. Alkali metals also have an irritating and caustic effect on the mucous membranes and skin.

The effect of barium compounds and strontium on the organism is characterized, in addition to the aforementioned, by the possibility of a general toxic effect [116, 284, 382, 453]. Acute poisoning is rare. Dyspeptic phenomena and changes on the part of the cardiovascular and nervous systems predominate among the symptoms.

The clinical picture of chronic poisoning by barium compounds includes affection of the air passages (coryza), inflammation of the oral mucosa, conjunctivitis, salivation, diarrhea, irregular or frequent pulse, dyspnea, high blood pressure, general weakness, difficult urination, and loss of hair. As a result of long work in contact with high concentrations of barium compounds, pneumoconiosis, often accompanied by acute inflammations of the lungs and bronchi, can develop [284, 382, 453].

The effect of magnesium is similar to that of barium and consists in irritation of the air passages and mucous membranes and in dyspeptic and nervous disorders. Gastric disease is accompanied by colic-like pains, sometimes with nausea and vomiting. Inhalation of magnesium fumes can be the cause of a combination of symptoms known as "metal fume fever." Metallic magnesium and its compounds, like other alkaline earth metals, cause skin lesions, irritation, dermatitides, and ulceration, which do not readily respond to treatment.

Little has been published on the effect of rare earth elements (lanthanides) on the human organism, apparently as a consequence of the small scale of their use in industry. Necrotic changes found in the liver of persons who died as a consequence of poisoning by these elements indicate the toxic effect of rare earth elements. The available experimental data [144, 277, 331, 450, 466] indicate the possibility of the occurrence of inflammatory and pneumoconiotic changes in the lung tissue and general toxic phenomena with affection of the parenchymatous organs, circulatory system, and skin.

The symptoms of affections in persons working with the so-called noble metals (elements of the copper group) – silver and gold – and also with metals of Group VIII (ruthenium, rhodium, palladium, osmium, iridium, and platinum) have been described inadequately. There are single indications that unique pigmentation of the skin occurs under the action of silver, gold salts, and osmium compounds. In the case of long contact with silver salts, general toxic phenomena (argyria) are observed which are characterized by affection of the liver, gastrointestinal tract, and other organs, and also by irritation of the mucous membranes of the air passages and sometimes of the eyes, accompanied by diminution of vision. In the case of protracted contact with gold and platinum compounds, allergic conditions are possible.

The data presented on the clinical manifestation of the toxic effect of various metals on the organism permit the conclusion that they can be divided into several groups according to the character of pathological changes effected.

Irritation of the skin and mucous membranes is most characteristic of alkali and alkaline earth metals. The effect of the transition metals of Groups IV–VIII is characterized by bronchitis, inflammatory and pneumoconiotic changes in the lung tissue, and general toxic effects with predominant affection of the gastrointestinal tract and parenchymatous organs and moderate disorders of the peripheral blood. Here the toxic effect of transition metals of Group IV, with the exception of iron, is more pronounced; copper, zinc, cadmium, aluminum, and antimony come close to them. The most toxic effect is displayed by mercury, lead, and manganese, which are characterized by predominant affection of the central nervous and cardiovascular systems and disorders of the activity of the gastrointestinal tract, liver, and kidneys. A special place is occupied by beryllium, which causes the development of a unique state with predominance of both a fibrogenic and a general toxic effect.

The transition metals of Group VI and the majority of their compounds effect changes primarily in the lung tissue and are characterized by the development of bronchitis, and inflammatory-proliferative reaction, and benign pneumoconiosis. Iron has a similar effect.

Allergic conditions ("metal fume fever," asthmoid bronchitis, dermatitides) are characteristic of the action of many of the metals considered, including those belonging to the second group (in our classification). The role of metals in the development of occupational bronchial asthma has been described in detail in [58].

CHAPTER IV

Toxic Effect of Some Metal Powders and Powder Metallurgy Mixtures

Among the articles manufactured by the powder metallurgy method the most widespread are antifriction, friction, sealing, and wear-resistant parts for friction pairs produced from iron–graphite base materials and widely used in general machine and instrument manufacture, ship and turbine building, in railroad transportation, in the aircraft industry, and in space engineering. Articles manufactured from copper base materials, particularly electrical contacts used in various branches of technology (ranging from the weak-current to the high-temperature industry), and also antifriction copper–graphite bushings for machinery make up a high proportion of all such products.

The effect on the organism of the dust of copper and reduced iron, and also of their metallic mixtures has received little attention in the literature. There are only individual reports [122, 124, 222] on studies of the combined toxic effect of various compositions of metal powders. They indicate the possibility of intensification or diminution of the pathological process under the combined action of mixtures of metals or metals with nonmetals.

Hygienic Evaluation of the Effect on the Organism of Copper Powder and Some Copper-Base Powder Metallurgy Mixtures

It was indicated in the preceding chapter that among workers exposed to copper dust there is a high morbidity with temporary disability and diseases of the respiratory organs, cardiovascular system, gastrointestinal tract, etc.

In addition, there are experimental data [103, 129, 229, 421] indicating that the dust of copper and copper ore in certain doses can cause poisoning with affection of the respiratory organs, liver and changes of the differential blood count and the albumin–globulin ratio. Intensified pneumoconiosis, observed under the action of dust in copper mines, is due to quartz dust. A small admixture of copper powder to quartz intensifies its pathogenic action, which is manifested in a more pronounced nodular form and in a general diffuse sclerotic reaction developing in lung tissue.

We carried out special experiments [42, 47] to investigate in more detail the effect on the organism of powdered copper and some of its powder metallurgy mixtures, determine more exactly the dependence of the toxic effect on the level of content in the air, and to determine safe concentrations.

Effect of Copper Powder on the Organism. For the investigations we used an electrolytic copper powder containing 98% Cu and having in its composition 80-95% of dust particles up to 4 μ.

In the first series of experiments powdered copper was administered endotracheally to 10 white rats. During the six months of the experiment two animals died and the condition of the others did not differ from that of the control animals. After six months the increase of collagen [322] in the lung tissues of the experimental animals in comparison with that of the controls was 41.0% and the content of ascorbic acid averaged 169.0 mg%, i.e., increased by 19.3 mg% in comparison with the control. The quantity of ascorbic acid in the blood averaged 0.29 mg% and did not differ from its content in the control animals.

Pathomorphological investigations revealed in the lungs focal thickening of the septa between alveoli (interalveolar septa), the peribronchial lymph nodes were enlarged, and an appreciable quantity of desquamated epithelial cells and dust particles of a dark-brown color was noted in the lumen of the large bronchi. The last were also found in the interalveolar septa and vascular walls. Lymphoid infiltrates were detected around the small and medium bronchi and vessels, and some alveoli were distended emphysematously. Moderate pathological disorders were noted in the parenchymatous organs; local swelling of the muscle fibers of the myocardium, liver cells, and epithelium of the renal convoluted tubules was observed. In the spleen we noted here and there atresia of the follicles and hyperplasia of the red pulp, which contained numerous brown bodies, probably hemosiderin. The zona fasciculata of the cortical layer contained a large quantity of lipids.

Three of four rabbits died as a result of a single inhalation of copper dust in a poisoning chamber at a concentration of 800-960 mg/m^3.

In the case of repeated inhalation poisoning of animals with copper dust in concentrations of 200-350 mg/m^3 (average 315) salivation and copious nasal discharge were noted in all rabbits during the first days, and two animals died. The condition of the survivors during the month-long experiment was distinguished by some torpidity, decrease of appetite, and intensified thirst.

A pathomorphological study of the internal organs revealed in the lungs swelling of the walls of the vessels and bronchi and perivascular and peribronchial infiltration by lymphoid elements. Desquamated epithelial cells and brown pigment bodies (dust) were noted in alveoli.

Fatty degeneration of the liver and parenchymatous degeneration of the epithelium of the renal tubules occurred in some animals.

The content of collagen in the lung tissue of rats sacrificed at the end of the experiment had increased by 50% in comparison with the control and the content of ascorbic acid by 359%. The content of ascorbic acid in the blood was the same as in the control animals. An investigation of the morphological composition of the blood of animals exposed to copper dust in a concentration fo 10-20 mg/m^3 showed a slight drop of the hemoglobin level, an increase of the number of segmented neutrophils, and a decrease of the number of lymphocytes; the number of erythrocytes, the total number of leukocytes, and the color index of the blood did not change.

The increase of collagen in this group of animals after a month was 65.6%, and then dropped slightly. The ascorbic acid content increased by 59.6% after a month and almost doubled after six months. A histological investigation revealed in the lung tissue small cellular-dust nodules in the center of which dust particles and small vessels with thickened walls were sometimes seen (Fig. 12). Desquamative bronchitis and peribronchial and perivascular

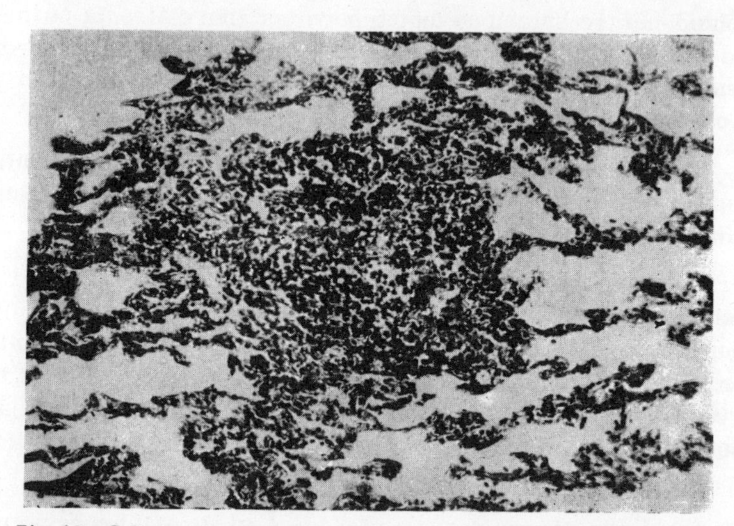

Fig. 12. Cellular-dust nodule in rat lungs six months after chronic exposure to copper dust in a concentration of 10-20 mg/m^3. Magnification 7 × 20.

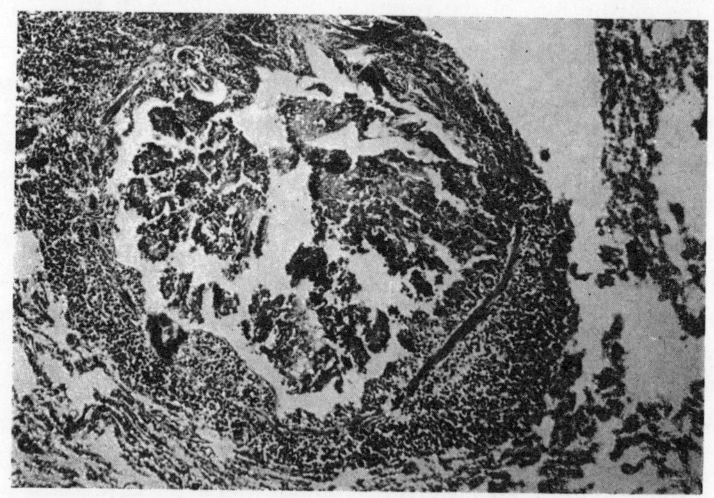

Fig. 13. Desquamative bronchitis three months after chronic exposure to copper dust in a concentration of 10-20 mg/m^3. Magnification 7 × 40.

lymphoid-cell infiltrations together with areas of emphysema were noted (Fig. 13). Changes in the parenchymatous organs did not differ substantially from those described earlier. Pronounced edema was observed in the submucosal layer of the stomach.

Some of the animals were subjected to 6 months inhalation poisoning by copper dust in a concentration of 1-10 mg/m^3 (average 6.7).

No changes were noted on the part of the morphological composition of the blood of the animals in this series. Nor was there an appreciable increase of collagen during the first four months, although the ascorbic acid level in the lung tissue reached high limits – it increased by 58.7-123.5%. An increase of the collagen content by 52.2% and of ascorbic acid by 32.8% was noted in the animals only six months after the start of the experiments.

A pathomorphological investigation of the internal organs after poisoning for one month revealed thickening of the bronchial walls, peribronchial lymphoid infiltrates, and accumulation of lymphoid elements in the interstitial tissue of the heart, and in some cases local swelling of the muscle fibers. Feeble local swelling of the liver cells and plethora of the vessels were noted in the liver, and swelling of some epithelial cells of the renal convoluted tubules was observed. A pink homogeneous mass was seen in many glomeruli of Bowman's capsule. Morphological peculiarities were not elicited in the adrenal glands, but plethora of the red pulp was noted in the spleen.

After longer periods of poisoning the changes were more pronounced, especially in the liver.

Taking into account that the disorders caused by copper dust are similar to the changes following inhalation and endotracheal administration of nickel dust [229] (for which the maximum allowable concentration is 0.5 mg/m^3), we consider it possible to recommend 0.5 mg/m^3 as the maximum allowable copper dust concentration.

Effect on the Organism of Some Powder Metallurgy Mixtures Produced in the Manufacture of Copper-Base Articles. To determine the effect on the organism of the dust produced in the manufacture of copper-base articles, mixtures of the composition most frequently used in production (70% Cu + 30% Ni; 90% Cu + 10% Sn; 97% Cu + 3% graphite) were administered endotracheally to animals. These mixtures contained 60-80% of particles up to 3 μ in size.

The pathological changes in the animals induced by these composites were compared with the changes effected by copper and cupric oxide powder of the same particle size analysis and with the condition of the control animals. The experiments were performed on 120 white rats which were observed for three and six months.

The investigations showed that, following the administration of various mixtures of copper dust with other powdered substances – nickel, tin, graphite – the changes in the organism are for the most part the same as those induced by powdered copper and its oxide.

The compositions investigated, like copper powder itself, caused a feeble fibrogenic response. The collagen content in the lung tissue increased by 16-32% in comparison with its level in the control animals.

The most active synthesis of collagen was induced by cupric oxide. After three months the increase of collagen reached 185% in comparison with the control.

The ascorbic acid level in the blood plasma did not change after administering these dusts, whereas in the lung tissue the ascorbic acid content decreased after administering the copper–tin mixture, and was at the control level after introducing the copper–graphite mixture. After administering the copper–nickel mixture, a slight increase was observed.

As in the case of administering copper dust, the mixtures induced thickening of the vascular walls even to their obliteration, and also of the bronchial walls due to proliferation of lymphoid-cell elements; the pulmonary interalveolar septa were as a rule thickened and connective tissue proliferated here and there along their course (Fig. 14). The last was especially pronounced following the administration of cupric oxide.

Six months after administering the copper–nickel mixture cellular-dust nodules were found in the lung tissue, whereas after the effect of the copper–tin mixture at this period only thickening

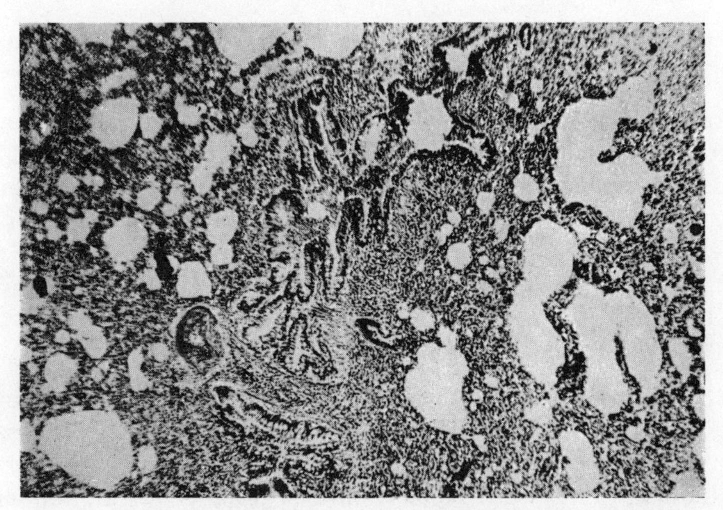

Fig. 14. Bronchopneumonia and thickening of vascular walls five months after endotracheal administration of copper-graphite powder. Magnification 7 × 8.

of the bronchial walls owing to proliferative infiltration around them was observed. The changes in the parenchymatous organs were also similar to those described earlier.

Conglomerates of dust particles were revealed in the lungs of the experimental animals even six months after the administration of the dust. This indicates slow removal of such dusts from the organism.

Since the changes in the organism induced by the compositions investigated are similar to the disorders effected by copper powder dust, for which we proposed a maximum allowable concentration of 0.5 mg/m^3, we consider it possible to recommend this concentration as the maximum allowable for the mixtures under consideration. For cupric oxide, the fibrogenic response to which is quite pronounced together with a general toxic effect, we consider it necessary to reduce the maximum allowable concentration by a factor of five, to 0.1 mg/m^3.

Toxic Effect of Dust of Iron Oxides and Iron Powders Produced by Reduction

The disease siderosis occurring upon inhalation of various iron oxides (arc-welding dust, scale), iron ores (siderosilicosis), etc., has been known for a long time. This is indicated by numerous clinical observations and experimental data.

Yet the problem of siderosis as an independent form of pneumoconiosis still remains disputable. Some authors (Deutsch et al.) regard siderosis as a pneumoconioses without fibrosis on the ground that the development of pneumoconiosis following inhalation of dust occurred benignly, and the roentgenologically detected nodular and focal changes were due to the roentgenographic contrast of the dust itself. Dvizhkov and others [91, 320] are of the opinion that iron dust is an independent etiologic factor of diseases.

They explain the roentgenologically observed "reverse development" of pneumoconiosis by the circumstance that, upon cessation of contact with iron-containing dust, the particles are gradually removed from the lungs as a consequence of phagocystosis, and feeble sclerotic changes remain at the site of dust nodules.

In addition to nodules, thickening of the interalveolar septa was observed in the lung tissue as a consequence of the deposition of dust particles and aggregation of dust cells. Conglomerates of dust particles with phenomena of diffuse sclerosis were detected in the lymph nodes.

Two types of siderosis are at present distinguished. Red siderosis occurs under the action of ferric oxide Fe_2O_3. The lungs and lymph nodes are enlarged, indurated, and in section contain a large number of dark-red polymorphous nodules, sometimes merging into large conglomerates which are penetrated by connective-tissue bundles. A microscopic examination reveals in the dust nodules many dust cells, freely lying dust particles, and lymphoid and histocytic elements. In many cases the alveoli situated around the nodules are filled with dust cells.

Black siderosis occurs under the action of ferrous oxide FeO, and also its carbonate and phosphate compounds. The lungs in this case acquire a black color as a consequence of the conversion of inhaled iron to siderin, which does not differ from the hemosiderin formed on disintegration of erythrocytes. The changes described were analogous to those observed in an experimental investigation of the effect of the ferric oxide dust in the form of a condensation and disintegration aerosol [211]. Similar phenomena were observed also by other authors [35, 64, 211, 241] who studied changes in the lungs effected by arc-welding dust, although its composition contained not only iron oxides but also the oxides of manganese and titanium, and other components. The quantitative determination of collagen in the lung tissue of white rats carried out in our laboratory [35, 241] at various times after the action of arc-welding dust containing predominantly iron oxides showed that the fibrogenic activity of this dust is one-fifth to one-sixth that effected by quartz dust under the same experimental conditions.

Pneumoconiotic changes in the lung tissue induced by various types of dusts of ferrous metallurgy plants containing predominantly iron oxides were studied in detail in our laboratory [12, 186] and by other authors [128]. The collagen level in the lung tissue of experimental animals after the administration of blast-furnace dust and dust of the mixer departments of the cast house of the blast-furnace plant did not exceed 30-40% of the collagen formation in the control animals, i.e., was one-fifth to one-sixth that observed under the action of silica dust. On the basis of these data the maximum allowable concentration (MAC) proposed for these types of dust was 6 mg/m^3 and for dusts containing iron oxide with an admixture of up to 10% silica, 4 mg/m^3. These MACs were approved by the USSR State Sanitary Inspection Office.

In addition to the changes described, some authors [175, 219] observed in persons exposed to iron-ore dust and arc-welding aerosols a high content of hemoglobin and erythrocytes in the peripheral blood. However, the results of investigations of other authors [15], who determined the effect of arc-welding dust on hematopoiesis, did not confirm these data.

The relatively low biological activity of oxide dusts can be due to the phagocytic reaction occurring in the organism after the uptake of iron-oxide dusts and directed toward rapid cleansing of the respiratory tract [12, 211]. Some investigators noted feeble degenerative changes in parenchymatous organs (liver and kidneys) [12, 106].

The effect on the organism of the dust of reduced pure iron has been studied much less than that of its oxides.

One author [402], who exposed guinea pigs to the dust of hematite (Fe_2O_3) and pure iron, did not find substantial changes after the administration of hematite, whereas pure iron caused mild fibrosis. In the opinion of another author [475], the dust of iron and its higher oxide does not cause pneumoconiosis. Doubt relative to the possibility of the development of pneumoconiosis under the action of pure iron dust is expressed also in [430]. In view of the divergence of the data on the biological effect of this type of dust, an experimental study of the fibrogenic and general toxic effect of the dust of pure iron was carried out under the supervision of Bakalinskaya [34, 42]. The experiments were conducted on 150 white rats observed for 9-12 months after the administration of the dust of iron powder.

The collagen content in the lung tissue, established by the method indicated earlier, was 56.4-80.6% higher in comparison with the control (the data are statistically significant). Maximum collagen formation (up to 80.6%) was noted three months after the administration of the dust, and after six and nine months it was about the same and amounted to 59.0-60.7%. If we take into account that under the influence of quartz dust the increase of collagen for the same experimental conditions is 250-300% [92, 322, 323], we can assume that the fibrogenic activity of iron dust is about 3-5 times less than the activity of silicogenic dust. The ascorbic acid content in the lung tissue of animals poisoned with iron powder after six months was almost the same as in the lung tissue of the control animals. Only after 12 months was its intensified biosynthesis observed, the content of ascorbic acid in the lung tissue showing an increase of 135.8%.

The ascorbic acid content in the blood of animals of both groups was about the same during the entire experiment.

A microscopic investigation of the internal organs of the experimental animals revealed during the first months a predominance of inflammatory-proliferative changes in the lung tissue and considerable accumulations of dust particles in the lumen of the alveoli, in the interalveolar septa, and in the bronchi.

At later periods (after 6 and 12 months) proliferation of the connective tissue around the bronchi and vessels and areas of emphysema in the interalveolar septa and lymph nodes were observed. In some animals cellular-dust nodules formed around the small vessels.

In the case of daily 4-h inhalation poisoning of animals with the dust of iron powders in V. B. Latyshkina's chamber for nine months at a concentration of 150-250 mg/m^3, no substantial changes were recorded in the behavior, body weight, and morphological composition of the blood. The collagen content during the entire period of the experiments increased on the average by 58-68% in comparison with its content in the lung tissue of the control animals. A noticeable increase of the biosynthesis of ascorbic acid in this case was observed only nine months after the start of the experiment.

The morphological changes in the lung tissue of the animals were similar to those described earlier and manifested themselves in the development of inflammatory-proliferative pneumoconiotic changes. Feeble degenerative disorders (parenchymatous degeneration) were observed in the parenchymatous organs (liver and kidneys).

In the case of inhalation poisoning of rats with dust of iron powder at a concentration of 1-10 mg/m^3 (average 6) for 12 months, no significant changes were detected in the organism.

Thus, the results of these experiments agree with the observations described above and convincingly demonstrate that the dust of pure iron powder can be the cause of pneumoconiotic changes. This is indicated not only by the results of morphological investigations but also by the data yielded by a quantitative determination of the fibrogenic effect of dust, i.e., the biosynthesis of collagen. Iron dust is an etiologic factor in the development of chronic bronchitis. The fibrogenic activity of this dust is weaker than that of silica by a factor of about 3-5.

Taking into account the similarity in the development of pathological phenomena under the action of the dust of iron powders and iron oxides, for which a maximum allowable concentration of 6 mg/m^3 has been established, we can recommend 6 mg/m^3 also as the MAC for the dust of reduced iron. A check of the effect of this dust on animals at this average concentration for 12 months confirmed the safety of the recommended MAC.

Pneumoconiotic Changes and General Toxic Effect of Carbonyl Iron Powders

Iron powder produced by the carbonyl method is distinguished not only by its high purity but also by its special onion-skin structure giving rise to the dielectric properties of its particles. Thanks to this, carbonyl iron powders are used widely in the manufacture of articles for electronic apparatus. Most frequently used in industry are high-purity carbonyl powders of grade V-3, in which impurities amount to $10^{-3}\%$; grade R-10 containing 99.8% elemental iron, magnetite, iron nitride, and cementite; and iron powder R-10, whose particles are coated with a film of SiO_2.

No information on the biological activity of the dust of various types of carbonyl iron has been found in the literature. Taking into account the small particle size of these powders and the unique structure of their particles, we can expect their biological action to be more pronounced than that of reduced iron powder.

In connection with this, the author of [44], together with coworkers from his laboratory, conducted investigations to determine the fibrogenic and general toxic effect of carbonyl iron powders.

The experiments were carried out on 300 white rats. The substances investigated (iron powder grades R-10, V-3, and SiO_2-coated R-10) were administered to the animals endotracheally: Some of the animals were exposed to inhalation poisoning in the chamber designed by V. B. Latushkina at concentrations of 150-200 and 1-10 mg/m^3 (average 6.0); the animals were observed after 1, 3, 6, and 9 months. Particles less than 4μ in size amounted to 91% in the dust of V-3 iron, 99% in R-10, and 59% in SiO_2-coated R-10 iron.

The observations showed that the condition of the experimental animals during the entire experimental period after endotracheal administration of the various grades of iron dust did not differ from the condition of the control animals.

The collagen content in the lungs of the animals three months after endotracheal administration of the powders of high-purity iron and grade R-10 increased maximally by 47-53% ($P < 0.05$). After the administration of SiO_2-coated R-10 iron collagen formation increased by 18-31% within one month, rose to 94% after six months, and thereafter remained at a high level.

The increase of ascorbic acid concentration in the lung tissue after the administration of V-3, R-10, and SiO_2-coated R-10 iron powders was statistically significant ($P < 0.05$) only after a sixth-month period. At other times, and also after the endotracheal administration of R-10 iron powder, the increase of ascorbic acid content was statistically insignificant. Nor were statistically significant changes in the content of nucleic acids (RNA and DNA) observed in the lung tissue after the administration of the test materials. A month after administration of V-3 iron foci of thickening of the interalveolar septa, primarily at sites of accumulation of dust particles, were noted in the lung tissue.

After three months perivascular and peribronchial infiltrates, hyperplasia of the peribronchial lymph nodes, and focal thickening of the interalveolar septa due to proliferation of the lymphoid-histiocytic elements and intensified exudation into the cavities of the alveoli together with areas of emphysema, developed. Single cellular-dust nodules were observed at places of aggregation of dust particles.

Subsequently, after six and nine months, the structure of the lungs in most cases did not change substantially.

In addition to disorders in the lung tissues, phenomena of cloudy swelling of varying degree were noted as early as three months after administration of the dust in the liver, heart, and kidneys, and in some cases fatty degeneration in the liver.

In stain tests on histological sections designed to reveal iron (Perls' test), the most intense blue color with a more intense dark hue at places of aggregation of dust particles was observed in the lung tissue in the first month after endotracheal administration of V-3 iron. Intense blue staining was observed also in the red pulp

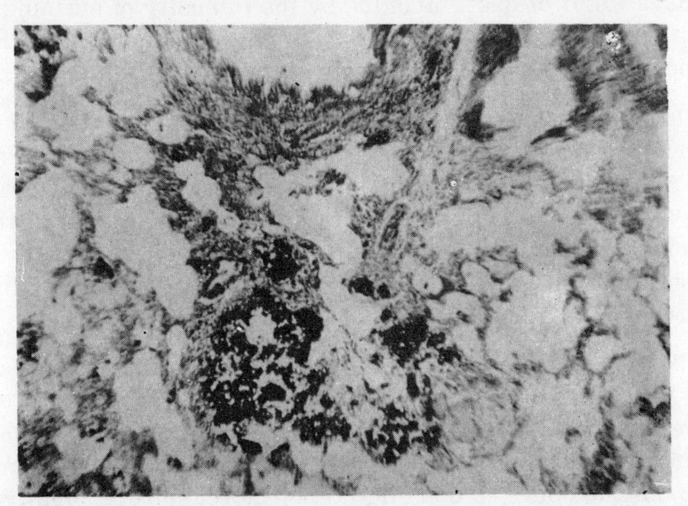

Fig. 15. Cellular-dust nodule and moderate fibrosis of lung tissue six months after endotracheal administration of carbonyl iron powders whose particles were coated with a SiO_2 film. Magnification 7 × 10.

of the spleen of the experimental animals, which was more pronounced in them than in the control animals.

Three months after the administration of V-3 iron a blue color was found in the lung tissue much more rarely than in the first months, and after six and nine months was completely absent, while remaining intense in the splenic red pulp.

After endotracheal administration of R-10 carbonyl iron similar pathological phenomena were detected, but degenerative changes were less pronounced.

In contrast to these experiments, in animals that received R-10 SiO_2-coated iron powder, the cellular-dust nodulues in the lung tissue appeared as early as one month after the administration of the dust. After three months the inflammatory process declined, and by the ninth month the number of cellular-dust nodules penetrated by a fine network of argyrophilic fibers also decreased (Fig. 15).

In this case the same pathological changes occurred in other organs as after the administration of carbonyl iron V-3 and R-10.

A comparison of the pathological changes in the lungs and other organs with the results of a histochemical study of the distribution of iron gives grounds to assume that the organism gradually cleanses itself of dust. Judging by the intensity of staining of sections of the spleen in tests for iron, we can suppose that the reticuloendothelial system participates considerably in this process.

Three and six months after inhalation of R-10 carbonyl iron in a concentration of 150-200 mg/m^2 the body weight of the experimental rats dropped and amounted to, respectively, 71-62% of the weight of the control animals ($P = 0.01\%$). Some of the animals died during the first two months of the experiment. No changes were noted in the behavior of the animals in the case of inhaling R-10 powder in concentrations of 1-10 mg/m^3.

The collagen content in the lung tissue of the animals exposed to R-10 iron powder in a concentration of 150-200 mg/m^3 three months after the start of the experiment had increased 54% (statistically significant) and after six months 79%. At concentrations of 1-10 mg/m^3 the collagen content did not increase ($P > 0.05$). Six months after poisoning with the dust of R-10 iron in a concentration of 150-200 mg/m^3 the ascorbic acid content had increased 103%.

In the other cases the levels of ascorbic and nucleic acids did not show significant statistical change. The morphological disorders in the lung tissue of animals exposed to R-10 iron were similar to those described earlier.

Similar but more pronounced phenomena were detected in the parenchymatous organs six months after the start of the experiment.

Hyperplasia of the peribronchial lymph nodes and feeble inflammatory changes in the lung tissue were noted following inhalation poisoning of the animals with R-10 iron in concentrations of 1-10 mg/m^3. Small cellular-dust nodules were observed here and there around the dust aggregates. Feeble phenomena of cloudy swelling and fatty degeneration were noted in the liver tissue and renal parenchyma.

The results of the observations indicate that the substances investigated cause a slight increase of the number of leukocytes in the peripheral blood, probably due to the peculiarities of the process of adaptation of the organism. Blood coagulability did not change in this case, whereas the level of lysozyme and complement was lower, which indicates a decrease of natural immunity of the organism. Restitution of the titer of complement under the action of R-10 SiO_2-coated powder was delayed in comparison with its restitution after the administration of powders V-3 and R-10. This indicates a more pronounced deleterious effect of powder whose particles are coated with SiO_2 in comparison with the effect of the other materials.

Investigations of the immunobiological activity of the organism also confirmed the untoward effect of R-10 powder in concentrations of 1-10 mg/m^3 (average 6.0); a decrease of the titer of lysozyme and complement was observed in the animals under the conditions indicated.

Summing up the data obtained in these experiments, we can conclude that the substances investigated have different fibrogenic actions. The fibrogenic activities of powders V-3 and R-10 are about the same and correspond to the effect of powders of reduced iron; the fibrogenic activity of R-10 SiO_2-coated powder is slightly higher. Special staining of sections of the lung tissue and spleen in tests for iron established self-cleansing of the organism of dust in which the reticuloendothelial system of the organism probably takes an active part.

Taking into account the foregoing, and also that at an iron powder concentration in the air of 6 mg/m^3 slight morphological and immunobiological changes were observed in the animals, we consider it necessary in this case to reduce the MAC in comparison with the MAC for dusts of iron oxide.

As the MAC for dust of carbonyl iron powders V-3 and R-10 we can recommend 4 mg/m^3 and for carbonyl iron R-10 whose particles are coated with a film of SiO_2, 2 mg/m^3.

Toxicity Characteristics of Some Powder Metallurgy Composites of Iron with Graphite and Nonferrous Metals

Powders of graphite and nickel in amounts of 5-40% and reduced chromium in amounts of 30-50% are often used (in the composition of mixtures) as alloying additions in the manufacture of machine parts. The introduction of these additions improves strength, wear resistance, and other properties [267, 351].

In view of the possibility of powder metallurgy workers being exposed to the dust of such powder mixtures, special investigations were carried out in our laboratory [34] to determine the effect on the organism of the principal powder charges used in the manufacture of iron-base articles containing 70-80% of particles up to 5 μ in size.

In the first series of experiments on white rats we investigated the biological activity of powder of an iron–graphite mixture (97% iron powder and 3% graphite) in comparison with the effect of reduced iron powder and graphite dust.

Six months after the administration of the graphite dust and iron–graphite powder the increase of collagen scarcely changed, amounting to 44.9-42.3%, whereas under the action of iron powder it increased to 66.4% in comparison with the control. The ascorbic acid level in the lung tissue of the experimental animals was about the same as for the control animals (20-22.8 mg%).

An examination of histological preparations of the organs of the experimental animals established that the morphological changes in the rats that received the graphite powder were characterized by the development of a feeble inflammatory-proliferative reaction in the lung tissue in the presence of a small content of dust particles. This attests to a phagocytic reaction to the administration of dust.

Similar changes were detected in the lungs and other organs of animals that had received the iron powder and iron–graphite mixture: Thickening of the walls of the vessels and bronchi, peribronchial and perivascular proliferation of lymphoid-histiocytic elements with sclerotic phenomena, and hyperplasia of the lymph nodes were observed. Feeble cloudy swelling was noted in the liver, heart, and kidneys.

Thus, when evaluating the pathological effect of an iron–graphite composition containing 3% graphite it is necessary to take into account the similarity between its biological effect and the effect of the dust of reduced iron. The MAC for such dust should be the same as for iron dust, 6 mg/m^3. We must assume that with an increase of the graphite content in the mixture the general toxic effect of the dust will be less pronounced, and therefore the MAC for such dust can be increased.

Following the endotracheal administration to white rats of an iron–nickel mixture (70% Fe and 30% Ni) and the powder of pure nickel containing 80% of particles up to 5μ in size, the behavior of the animals was characterized by some adynamia. A slight lagging of weight gain was noted in the rats that had received nickel powder.

The increase of collagen in the animals exposed to pure iron after six months was 66.4% and after administration of the iron–nickel mixture 71.2%, whereas after the administration of nickel the increase of collagen was 131.7%. The ascorbic acid level in the lung tissue under the effect of the nickel dust and iron–nickel mixture decreased to 65.9 and 74.5%, respectively, in comparison with the control.

Hence, the data obtained indicate a high fibrogenic activity of the iron–nickel mixture, which differs considerably from the activity occurring after exposure to nickel dust and is closer to the effect of iron dust. The decrease of the ascorbic acid content in the lung tissue under the action of the iron–nickel mixture can be regarded as a result of its increased use in the biosynthesis of collagen. The role of ascorbic acid in the organization of collagen fibers is known [99, 216, 259]. Thus the elevated content of ascorbic acid indirectly indicates a higher fibrogenic activity of the mixture in comparison with the activity of iron powder, which is probably due to the effect of the nickel component.

The morphological changes in the animals following exposure to both types of dust were similar. Marked inflammatory changes in the lung tissue, atelectasis and emphysema, pronounced thickening of the walls of the vessels and bronchi with perivascular and peribronchial infiltrates from the lymphoid-histiocytic elements were observed. Phenomena of cloudy swelling developed in the heart, liver, and kidneys which were considerably more pronounced than after the administration of iron powder.

The data presented above on the changes in animals exposed to nickel dust and iron–nickel powder are consistent with Mogilevskaya's data [226]. However, the changes occurring under the action of the dust of an iron–nickel alloy containing 30% Ni, as described by that author, differ from those observed in our work by a feeble toxic effect. This can be attributed to the circumstance that, in the alloy, the nickel was less free (as a result of interaction with the iron) than in the mixture, where it could act as an independent component.

Taking into account the foregoing, we must assume that in a hygienic standardization and evaluation of the biological aggressiveness of dust under industrial conditions it is necessary to be guided by the content of nickel in its composition with iron. With 30% nickel in the mixture we can recommend on the basis of the MAC established for it (0.5 mg/m^3) that the MAC for the dust of the iron–nickel composition should be 1 mg/m^3.

In the case of endotracheal administration of an iron–chromium mixture containing 30% of chromic oxide powder the increment of collagen in the lung tissue of the experimental animals that received this mixture was approximately the same as after administration of iron dust (66.4%), i.e., increased 53.2%; the ascorbic acid content increased 43.6%.

A morphological investigation of the internal organs of the experimental animals revealed not only phenomena of bronchitis but also marked irritation of the mucous membrane of the bronchi and in some cases its ulceration. After exposure to this dust marked proliferation was observed around the vessels and bronchi and along the course of the interalveolar septa. The walls of many vessels were thickened, sometimes with obliteration of the lumina. Thickening of the interalveolar septa was noted in places, the lumina of the alveoli were filled with an inflammatory exudate containing desquamated alveolar epithelial cells. Emphysema of the lungs was noted in certain places. Together with this there occurred swelling and focal parenchymatous degeneration of the myocardium, liver, and epithelium of the renal convoluted tubules as well as hyperplasia of the splenic red pulp, in which numerous hemosiderocytes could be seen.

Thus the pathomorphological changes in the internal organs and the fibrogenic action of the dust of iron–chromium powder were more pronounced than those of iron dust. When establishing the MAC for an iron–chromium mixture it is necessary to be guided by the content of chromic oxide powder in it.

Thus, if a composite contains a small amount of a low-activity substance, for example, graphite, its pathogenic activity is due to the toxic properties of the principal (in quantity) component. When a mixture contains toxic materials the general effect is determined by their content and consequently the MAC for the mixture must be established on the basis of the quantity and allowable concentration of the most toxic ingredient.

Fibrogenic and General Toxic Effect of Some Powdered Ferrite Composites

Among sintered articles used in electronics and electrical engineering, ferrites of various grades are gaining importance. Magnesium–manganese, manganese–zinc, nickel–zinc, lithium–zinc, chromium-containing, yttrium, and other ferrites have become commonplace. Their main components are oxides of iron (Fe_2O_3), manganese, yttrium, and other metals. However, the effect on the organism of the dust of mixtures used for manufacturing various ferrites has been examined only in single studies.

Kovalevich et al. [132] investigated the dust of the charge and pulverized ferrite M-18 containing 70.4% Fe_2O_3, 15.8% MgO, 11.6% Mn_2O_5, and 2.2% NiO.

Differing data were obtained following endotracheal administration of the indicated materials. Some of the animals that had received the ferrite charge died within 2-9 days from bronchopneumonia and pulmonary edema. In animals sacrificed after one and three months feeble catarrhal bronchitis, a weak interstitial proliferative process with a high content of polymorphonuclear leukocytes in the septa, and microfocal emphysema were noted. Compact, diffusely situated cellular-dust nodules – granulomas, consisting mainly of giant cells of the foreign-body type – were observed in the lung tissue. The nodules gradually diminished. Collagen fibers appeared as the dust disappeared, but the development of the fibers around the nodules, bronchi, vessels, and in the interalveolar septa was not pronounced; even 15 months after the administration of the charge a feeble sclerotic effect was observed. The bifurcate lymph nodes were hyperplastic; they contained numerous dust particles, giving, as did the particles of the granuloma, a positive reaction for iron.

The dust of the finished ferrite has a similar effect, but less pronounced. Unlike the effect of the charge, in this case proliferative phenomena predominated, whereas the mixture of oxides effected primarily an inflammatory process.

Investigations in our laboratory [34] showed that, following the endotracheal administration to white rats of a magnesium–manganese ferrite charge containing 43% Fe_2O_3, 28% MgO, 19% Mn_2O_5, and 5% each of ZnO and CaO, feeble sclerotic changes occurred in the lung tissue. The collagen content in it after three and six months in animals that had received the ferrite charge increased 59% in comparison with the control and under the action of MgO increased respectively 50.6% and 82.9%, i.e., the increase of collagen was even slightly higher.

The administration of manganese oxides was accompanied by lower collagen formation, the level of which was 43.6-21.2% with respect to the control. In the case of inhalation of the dust of ferrites in concentrations of 100-200 and 1-10 mg/m^3 we also observed an increase of the collagen content; for the high dust content it increased 71.5% after six months.

Thus the fibrogenic activity of the charge investigated was a quarter to one-fifth times that of quartz, under whose action the increase of collagen reached 250-300% [92, 323], and about the same as for the dust of iron oxides and pure iron. However, the general toxic effect of the composite investigated was more pronounced than that of the iron oxides. Phenomena of desquamative bronchopneumonia with appreciable areas of atelectasis and thickening of the interalveolar septa together with foci of emphysema were observed in the lung tissue, mainly at the sites of dust aggregates. Feeble tissue sclerosis was noted later. The walls of the bronchi were thickened owing to proliferation of lymphoid-cell elements and histiocytes, and as a rule desquamated epithelial cells were found in the bronchial lumen. The walls of the small vessels were thickened, the lumina constricted, and sometimes completely obliterated. Lymphoid-cell infiltration with the later development of feeble sclerosis was noted around the vessels and bronchi.

In the parenchymatous organs swelling of the cells and phenomena of focal cloudy swelling (in the myocardium, liver, and epithelial cells of the renal convoluted tubules) were observed at the start of the experiment. Fatty degeneration of the liver was noted at later periods (after six months).

When determining the hygienic standard of ferrite dust it is apparently necessary to take into account the content of the most toxic components causing the principal pathological changes. In view of this, 1.5 mg/m^3 is probably the maximum allowable concentration for dust of magnesium–manganese ferrite composites containing 19% of manganese oxides.

A feeble fibrogenic reaction and inflammatory-proliferative changes in the lung tissue and bronchi with hyperplasia of the bifurcate lymph nodes were observed following endotracheal administration to animals (white rats) of a manganese–zinc ferrite charge containing 27.6% Mn_2O_5, 3.5% ZnO, and 68.9% Fe_2O_3.

Insignificant degenerative changes were noted in the parenchymatous organs. Consequently, toxic phenomena also predominate following exposure to such a mixture, and one must therefore be guided first of all by the content of manganese oxides when determining the MAC.

The charge of nickel-zinc ferrites containing 50% Fe_2O_3, 25% NiO, and 25% ZnO induced in the animals less appreciable changes in comparison with those described above and were characterized by an inflammatory-proliferative reaction in the lung tissue analogous to that described. Cellular-dust nodules containing dust particles, dust cells, and lymphoid-histiocytic elements were observed at sites of dust aggregates. Feeble focal diffuse sclerosis of the lung tissue, around the vessels and bronchi, and of the peribronchial lymph nodes was noted at later periods. Feeble basophilia was detected in the parenchymatous organs of white rats exposed to this dust.

In connection with the foregoing, in determining the maximum allowable concentration of nickel–zinc ferrite dusts it is recommended that the calculation be based on the nickel content.

Administration of barium ferrite dust to animals was followed by death of some of them. An inflammatory-proliferative reaction along the course of the interalveolar septa, around the vessels and bronchi, mainly at sites of aggregation of dust particles, was noted in the lung tissue of the animals that survived. Phenomena of catarrhal bronchitis were also observed.

An investigation of histological preparations of the liver showed phenomena of cloudy swelling and sometimes, at later periods, fatty degeneration. A comparison of the changes observed with the data on the effect of barium on the organism [284, 285] permits the assumption that these alterations are due largely to its effect.

A comparison of the pathological disorders described with the changes induced by powders of other ferrites indicates a less pronounced toxic effect of barium ferrite charges.

On the basis of the data presented we can conclude that in the ferrite industry the dust of the charge, i.e., the mechanical mixture of various metal oxides, is the most dangerous. Despite the fact that such a charge contains predominantly ferric oxide, which effects mainly a feeble inflammatory-proliferative reaction with moderate sclerosis in the lung tissue, a toxic effect is characteristic of ferrite dust. The extent of the toxic phenomena is largely determined by the quantitative composition of manganese, nickel, etc. In connection with this, one must proceed from the content of these metals in the compostion when making a hygienic assessment of the possible pathological effect of ferrite dust and when determining its allowable concentration in the air of the working zone.

CHAPTER V

Toxic Effect of Refractory Compounds

Modern technology imposes high requirements on the physicochemical and processing properties of metallic materials. In connection with this, industry is employing more and more substances of new types, among which a significant place is occupied by refractory compounds having high hardness or superhardness, good heat and corrosion resistance, and low density [292, 295].

In powder metallurgy powdered refractory compounds are synthesized and various parts are manufactured from them.

The data presented earlier on the state of health of workers at hard metal plants and the characteristics of the working conditions in the synthesis of refractory compounds and manufacture of articles from them indicate the possibility of the occurrence of pathological states in persons coming into contact with these substances. The results of experimental investigations also indicate the deleterious effect of refractory compounds on workers.

Hygienic Evaluation of the Effect of Transition Metal Borides and Carbides on the Organism

Clinical observations of the state of health of persons exposed to borides have not been described in the literature. The character of the structure of boron, its crystal structure, and physicochemical properties indicate that it is an active element and apparently even in compounds can cause the development of pathological alterations in the organism. The activity of boron depends considerably on its degree of purity and crystallochemical properties [245].

Amorphous boron, which is rapidly oxidized in contact with air in the presence of moisture, deserves particular attention. It has pronounced hydrophilic properties, forming a stable sol in water. The latter is related with surface oxidation of boron particles upon their dispersion in water [245]. A high activity of boron following its action on the organism is confirmed by experimental data obtained in studies of the effects of boron, boric oxide, and boric acid [25, 127, 398, 545] on the organism.

Boron as a trace element is important for the conversion and utilization of carbohydrates, and also participates in the respiratory process. On entering the organism, it is deposited primarily in the bones. When boric acid enters the organism boron is found predominantly in the brain and liver tissues and in the adipose tissue.

Data on the effect of boric acid on the activity of a number of enzymes, vitamins, and hormones have been obtained in [60]. On the whole the toxicity of boron and its compounds is evaluated as moderate [545]. The results of more thorough experiments designed to reveal the toxic effect of boron and boric oxide are presented in [127]; the purpose of this investigation was to determine the toxic effect of powders of the crystalline oxide and elemental boron of different degrees of purity. Lethal doses of boric oxide, established for mice, are 1.8 g/kg with intraperitoneal injection and 3.2 g/kg with oral administration. High-purity elemental boron did not kill the animals, while the medial lethal dose of less purified boron was 6.4 g/kg.

Changes of carbohydrate and protein metabolism and dysfunction of the liver and nervous system occur under the action of boron and boric oxide.

In animals exposed to brief inhalation poisoning by boric oxide in a concentration of 150-230 mg/m^3, thickening of the interalveolar septa occurred in the lung tissue owing to infiltration by histiocytes, lymphoid cells, and a few leukocytes and fibroblasts. The peribronchial lymph nodes were hyperplastic.

Boron nitride effected perivascular infiltration. On the whole the alteration bore the character of a moderate, predominantly proliferative interstitial process. Substantial changes were not noted in other organs of the animals, which was apparently due to the brevity of the action. In longer experiments with oral administration of the substances feeble degenerative changes were noted in the myocardium, liver, and epithelium of the renal convoluted tubules.

On the basis of the results described of the investigations it is concluded that boron and its compounds have moderate toxicity; it manifests itself more strongly with the oxide. The recommended maximum allowable concentration for the dust of crystalline boron is 5-7 mg/m^3 [127].

It must be borne in mind that the brevity of the observations of the conditions of the animals following inhalation of boron does not give a clear idea about its effect on the organism.

This gap is filled to some extent by the data obtained in our experiments on white rats with endotracheal administration of amorphous boron [40, 46].

We established that amorphous boron produces a high fibrogenic reaction.

After 20 white rats were administered powdered amorphous boron containing 86% of particles measuring up to 2 μ, the condition of the animals and their body weight during the six months of observation were the same as for the animals of the control group. The content of collagen in the lung tissue after three months had increased by only 36% ($P < 0.05$). However, intensified biosynthesis of ascorbic acid was noted after this time (Table 18). Six

TABLE 18. Content of Collagen and Ascorbic Acid in the Lung Tissue of Rats after Administration of Amorphous Boron

Substance administered	Exptl. period, months	Collagen			Ascorbic acid		
		Absolute content	Increment		Absolute content	Increment	
		mg	mg	%	mg	mg	%
Amorphous boron	3	30.2±1.3	8.0±0.4	36.0	24.5±1.2	6.3±1.3*	34.6
	6	51.35±4.0	32.42±1.3	171.6	18.1±7.6	0.20±0.1*	1.1
Physiological solution (control)	3	22.2	—	—	18.2±3.1	—	—
	6	18.9±4.67	—	—	17.9±4.6	—	—

*$P > 0.05$; in other cases $P < 0.05$.

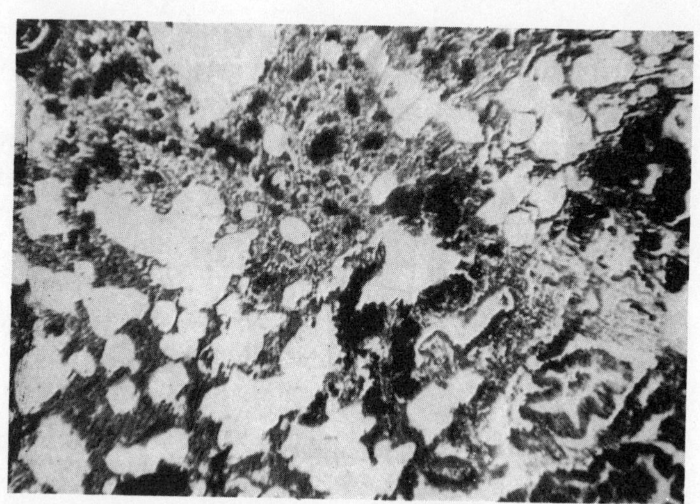

Fig. 16. Diffuse thickening of the interalveolar septa at sites of localization of amorphous boron particles three months after endotracheal administration of dust. Magnification 7 × 10.

months after the administration of boron the accumulation of collagen in the lung tissue was 171.6% higher than in the control animals.

Histological investigations revealed in some parts of the lungs aggregation of appreciable quantities of dust in the interalveolar septa, alveoli, and in the peribronchial lymph nodes. Thickening of the interalveolar septa due to proliferation of lymphoid-histiocytic elements and hyperplasia of the lymphoid tissue in the peribronchial lymph nodes were noted here. Small cellular-dust foci consisting of dust particles and dust cells, and also of lymphoid-histiocytic elements formed at the sites of dust-particle aggregates (Fig. 16). In addition, areas of emphysema were detected.

At later periods (six months) collagen fibers appeared in the dust nodules. Moderate diffuse fibrosis was observed along the interalveolar septa, in the peribronchial lymph nodes, and around the bronchi. Mucus, dust particles, and desquamated epithelial cells were found in the lumen of the bronchi (Figs. 17 and 18). There were no changes in the other organs (heart, liver, kidneys, adrenal glands, spleen).

Thus amorphous boron was found to induce a pneumoconiotic process and catarrhal bronchitis.

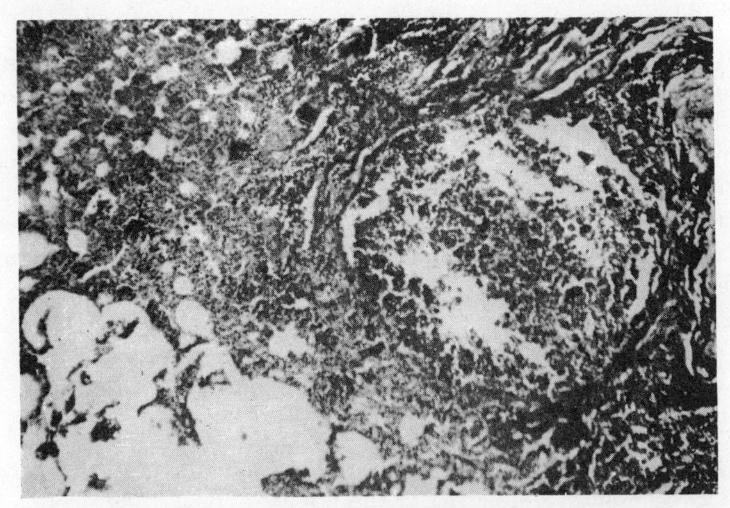

Fig. 17. Granulomatous nodules and bronchopneumonia six months after endotracheal administration of amorphous boron. Magnification 7 × 10.

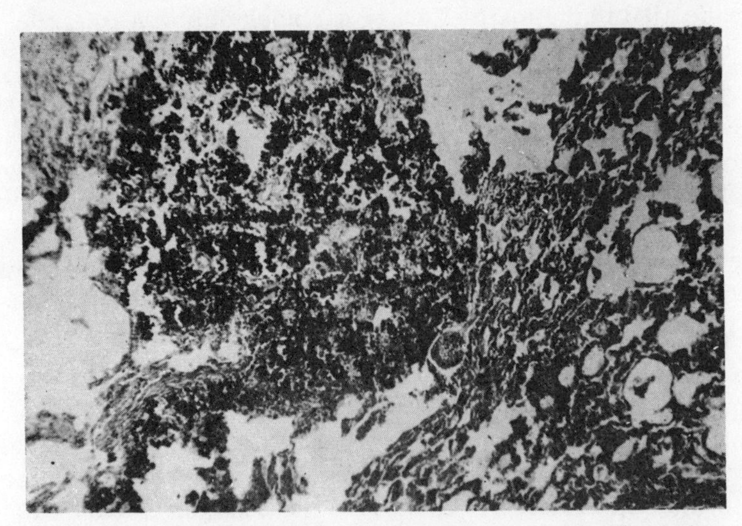

Fig. 18. Cellular-dust nodule in lung tissue three months after endotracheal administration of amorphous boron. Magnification 7 × 10.

The results of some investigations attest to the possibility of an injurious effect of carbides and borides on the organism.

Chromium boride and carbide effected an insignificant decrease of the hemoglobin content, moderate proliferation of lymphoid-histiocytic elements, and bronchitic and peribronchitic phenomena with a clearcut predominance of vascular alterations. Self-cleansing of the lungs and the formation of small dust nodules around the remaining dust particles with their subsequent encapsulation occurred at later periods. The content of serum albumins changed, together with a relative increase of the quantity of globulin fractions under the action of both types of dust. A lesser toxicity of the dusts investigated in comparison with the effect of chromic oxide and chromic anhydride was established on the basis of the data presented [282].

To obtain comparative data on the pathological effect of refractory compounds belonging to the boride and carbide groups, we conducted experimental investigations (on white rats) into the toxic properties of the borides of titanium TiB_2, zirconium ZrB_2, and chromium Cr_3B_2 in comparison with the toxicity of amorphous boron and metallic components of borides. A comparative study was made also of the toxic effect of carbides of boron B_4C, titanium TiC, zirconium ZrC, and chromium Cr_3C_2 [46].

The compounds in question were administered to the animals endotracheally in an amount of 50 mg/ml of sterile physiological solution. Observations of the content of collagen and ascorbic acid in the lung tissue and of the morphological alterations in internal organs (lungs, heart, liver, kidneys, adrenal glands, spleen) were carried out for one, three, and six months, and in some cases eight and 12 months after the administration of the substances.

The results of the investigations showed that metal borides and carbides have different fibrogenic effects. The borides of titanium, zirconium, and chromium, and also amorphous boron and boron carbide have a stronger effect compared with the carbides of titanium, zirconium, and chromium. Whereas three months after the administration of the borides the collagen content was 128-237% in comparison with the control, after the administration of zirconium and chromium carbides its content in the lung tissue increased by only 18 and 36%, respectively. The administration of titanium carbide did not affect collagen formation.

It can be assumed on the basis of these data on the effect of boron on the organism that the high fibrogenic action of borides is due to the effect of boron, which has a pronounced fibrogenic activity.

The peak on the curve of collagen formation under the action of the aforementioned substances was noted after three and six months; later, after eight and 12 months, it was close to the initial level.

The morphological changes in the internal organs of the animals a month after the administration of titanium boride were characterized by the appearance in the lung tissue of thickening of the interalveolar septa, chiefly around dust particles located in the cavity of the alveoli and interalveolar septa. After six months lymphoid-cell granulomas and proliferation of connective-tissue elements along the periphery of the dust nodules were observed together with the changes indicated. The number of dust particles in the lung tissue in this case was much smaller than after a month. In some animals pulmonary emphysema phenomena, a homogeneous albuminous mass around the pulmonary vessels, and lymphoid infiltration were noted over extensive areas.

Of the other organs the most affected was the liver, in whose tissue fatty degeneration phenomena were observed as early as one and three months after administration of the dust.

The development of the pathological process became more intense with prolongation of the period of the experiment. Three months after the administration of titanium boride the foci of fatty degeneration acquired appreciable dimensions (Fig. 19). The liver cells had vacuoles, and in some cases their boundaries and nuclei were feebly differentiated. In other liver cells the nuclei were in a state of pycnosis or were absent; a state of mitosis or binuclear condition, i.e., phenomena of intense regeneration of the liver tissue, was also noted.

No substantial changes occurred in the parenchyma of the kidneys. A large quantity of lipids was found in the cortical and medullary layers of the adrenal glands. Hyperplasia of some follicles was observed in the spleen.

After six months the phenomena described were more pronounced. Especially marked changes occurred in the liver tissue,

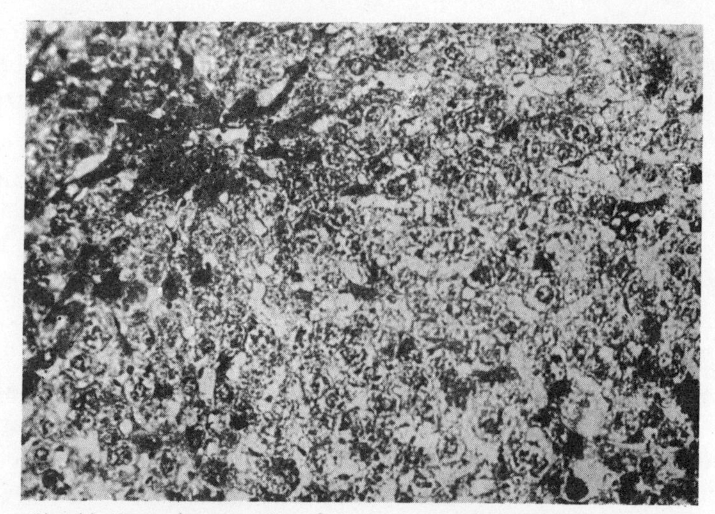

Fig. 19. Fatty degeneration of liver three months after endotracheal administration of titanium boride. Magnification 7 × 20.

where there were more extensive necrotic foci, whereas regenerative processes were manifested to a lesser degree.

One and three months after endotracheal administration of zirconium boride to white rats enlargement of the peribronchial lymph nodes was observed in the lung tissue. Single, black dust particles were found in the cavity of the alveoli after three months.

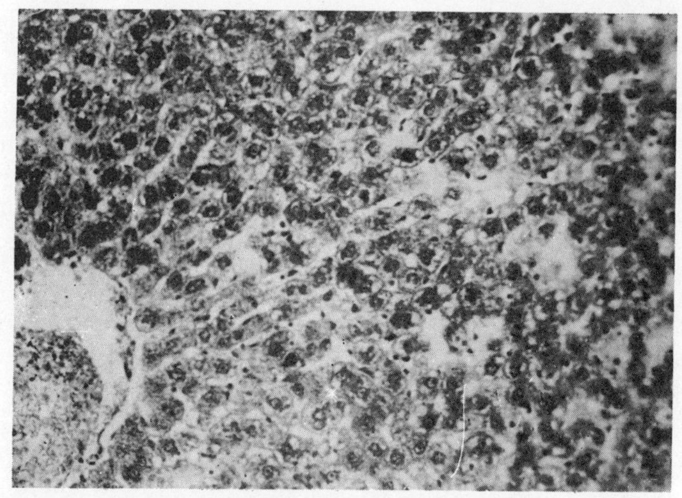

Fig. 20. Fatty degeneration of liver three months after endotracheal administration of zirconium boride. Magnification 7 × 20.

After six months proliferation of small nuclear cells developed along the bronchi, with atelectasis and initial stages of proliferation of the connective tissue around the vessels and small bronchi. Desquamation of the bronchial epithelium with marked discharge of mucus was noted.

Three months after the administration of zirconium boride phenomena of fatty degeneration and cloudy swelling developed in the liver, and dilatation of vessels containing a large amount of a dark-brown pigment was noted (Fig. 20). Many liver cells contained vacuoles with disintegrated cell nuclei. The boundaries and nuclei of the cells in some areas were feebly differentiated and the protoplasm was granular and swollen. After six months the phenomena of cloudy swelling and fatty degeneration were more pronounced. Necrobiotic foci stood out sharply in the centers of the hepatic lobules, while signs of regeneration were elicited only along the peripheries of the lobules. In certain instances necrobiosis was attended by the presence of "hematoxylin spheres." A large quantity of necrotic masses was noted in such foci together with inhibition or absence of regeneration. Phenomena of degeneration of the epithelium of the renal convoluted tubules with areas of necrosis developed in the kidneys under the action of zirconium boride (Fig. 21). No substantial alterations occurred in other organs.

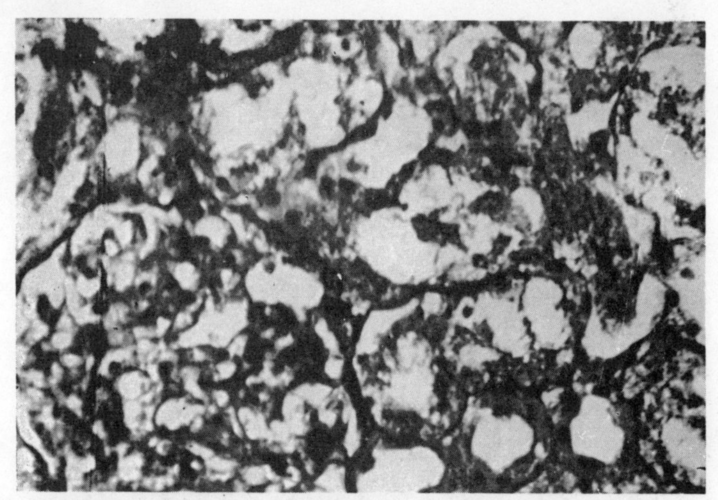

Fig. 21. Dilatation of the glomerular capillaries and degeneration of the epithelium of the renal convoluted tubules with necrotic areas six months after endotracheal administration of zirconium boride. Magnification 7 × 20.

A comparison of these results of investigations with data on the effect of zirconium oxide on the organism indicates the appearance of similar pathological phenomena.

A month after endotracheal administration of chromium boride to white rats the lung tissue contained many dust particles, but after six months only single dust particles were seen. Infiltration from the lymphoid cells and edema were found around some vessels, and the interalveolar septa in some places were thickened. Emphysematous foci occurred together with inflammatory changes.

Nonuniform basophilia of the muscle fibers and protoplasm of the epithelium of the capillaries was observed in the heart after 6 months. The muscle fibers were hyperplastic.

The most substantial changes occurred in the liver, where already after three months pronounced fatty degeneration with phenomena of suppression of regeneration occurred. In some cases there were necrotic foci with complete absence of regenerative phenomena. Metachromasia, foci of disintegration of nuclei, and their division were observed on liver preparations. Binuclear and multinuclear cells were encountered (Fig. 22).

Degenerative changes of the epithelium of the renal convoluted tubules occurred. The cavity of the glomerular capsules contained an albuminous exudate (Fig. 23).

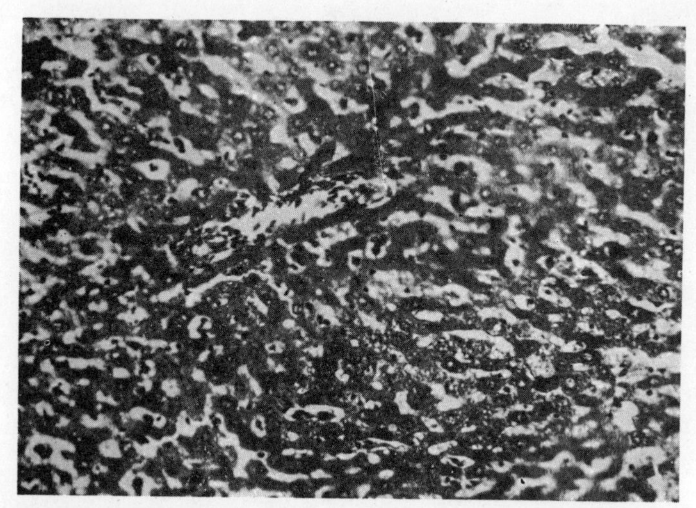

Fig. 22. Fatty degeneration of liver six months after endotracheal administration of chromium boride. Magnification 7 × 20.

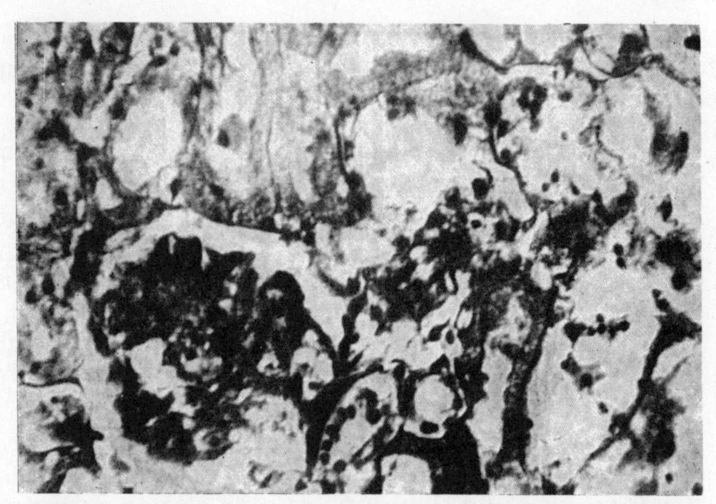

Fig. 23. Parenchymatous degeneration with necrotic areas of the epithelium of the renal convoluted tubules six months after endotracheal administration of chromium boride. Magnification 7 × 20.

The changes observed following endotracheal administration of chromic oxide were similar to those described, but were far more pronounced. Exudative bronchopneumonia, peribronchitis, perivasculitis, and perivascular pneumosclerosis developed in the lung tissue of the animals. Diffusely disseminated dust particles were found in the stroma of the lungs, and lymphoid granulomas were noted around the foci of dust aggregates. Changes of a degenerative and reactive character, analogous to those that occurred following the administration of chromium boride, occurred also in the liver, heart, and kidneys.

The results of these investigations show that the borides of titanium, zirconium, and chromium have a fibrogenic effect, in conjunction with which degenerative changes occur in the liver and kidneys under the action of these compounds, and after the administration of chromium compounds changes of a degenerative character occur also in the myocardium.

The compounds investigated can be arranged in the following order of decreasing toxic effect: chromium boride, titanium boride, and zirconium boride.

The morphological changes in the lung tissue induced by boron carbide are similar to those observed after the administration of amorphous boron. Some bronchi are thick-walled. Their lumina contain an exudate with a considerable number of desquamated epithelial cells and lymphoid cells. Considerable thickening of the walls of the small vessels with obliteration of some of their lumina is observed after longer periods (12 months). Phenomena of peribronchitis and perivasculitis are noted. The stroma of the lung contains single dust particles in the form of dark-colored granules around which thickening of the interalveolar septa, atelectasis, and emphysema occur.

The morphological changes induced by the administration of titanium, zirconium, and chromium carbides resembled the pathological state effected by the oxides of these metals. Diffuse thickening of the interalveolar septa, especially around the foci of dust-particle aggregates, was observed in the lung tissue (Fig. 24). Infiltration from lymphoid elements was seen clearly around the bronchi and vessels. Peribronchial sclerosis and pneumosclerosis were noted at later periods (after 12 months).

The morphological alterations in the liver, kidneys, and adrenal glands, expressed in varying degrees, indicate disturbance of metabolic processes in the organism.

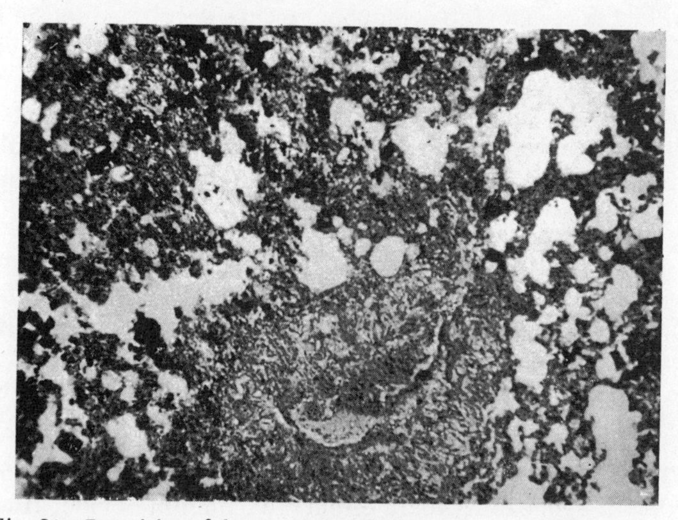

Fig. 24. Deposition of dust particles, thickening of interalveolar septa, and perivascular edema in rat lung three months after endotracheal administration of chromium carbide. Magnification 8 × 10.

The changes induced by the administration of the aforementioned metals were less appreciable than those effected by their oxides and borides. Titanium and zirconium carbides induced only phenomena of basophilia in the liver and kidneys, the vessels being affected less, while after the administration of chromium carbide no changes were found in the myocardium.

Comparing the disorders induced by metal carbides, we can conclude that their activity changes in the same manner as the activity of the borides, i.e., the toxicity of these compounds decreases from chromium carbide to titanium and zirconium carbide.

The refractory compounds effect degenerative changes in the parenchymatous organs, mainly in the liver, which is characterized by a protein-forming function. The changes noted on the part of serum protein, particularly the decrease of the quantity of albumins accompanied by a relative increase of the content of the globulin fractions under the action of chromium boride and carbide, are apparently related with dysfunction of the liver [282].

A decrease of immunobiological reactivity increases the level of morbidity among workers [240, 360, 361]. In our laboratory we carried out experiments designed to reveal the effect of chromium boride and carbide on some indices of natural immunity [443]. In experiments on white rats we studied the changes of the content of lysozyme and complement in the blood serum for six months after endotracheal administration of the materials. The results of these investigations indicate that the content of complement in the serum was low already after a month under the action of reduced chromium, chromic oxide, and chromium carbide. The minimum content of complement (as low as 27.6%) was noted after the administration of chromic oxide, whereas the effect of reduced chromium and chromium carbide was about the same (51.4 and 42.5%). Subsequently the content of complement increased, but the original level was not reached until six months later.

The titer of lysozyme in the serum dropped simultaneously with the decrease of the content of complement.

On the basis of these data we can conclude that the resistance of the organism is maximally weakened under the action of chromic oxide; it decreases to about the same extent after the administration of reduced chromium and chromium carbide.

The content of hemoglobin in the blood of the animals during the entire experiment did not differ appreciably from the original level and from its level in the control animals. The content of leukocytes did not change substantially either.

The time of the start of blood clotting under the action of chromium and its compounds was markedly increased as early as after a month. After the administration of the materials the prothrombin index of the animals decreased by almost one-half, then recovered, but did not reach the original level even by the sixth month.

In animals poisoned by the dust of chromic oxide and chromium carbide the collagen content in the lung tissue increased, and morphological changes identical to those described above occurred in the internal organs of the animals. On the basis of these experimental data we can conclude that the toxicity of chromium carbide is close to that of reduced chromium and much lower than that of chromic oxide.

Our data on the effect of refractory borides and carbides on the organism are in agreement with literature data.

Before the start of our investigations (1961-1962) there were no data in the literature on the effect of borides on the organism. There were only single reports on the pathological effect of titanium and tungsten carbides [199, 200]. At present we know of several studies on change of biological activity caused by molybdenum boride Mo_2B_5 [232], calcium boride CaB_6, titanium boride TiB_2, niobium boride NbB_2 [126], zirconium boride ZrB_2, and zirconium carbide ZrC [234], molybdenum carbide MoC [233], chromium carbide Cr_3C_4, and chromium boride CrB_2 [282], in which the different toxicity of these refractory compounds, which is probably related with the character of the internal structure and chemical activity of the substances, is indicated.

Calcium boride has the strongest toxic effect, for which the following lethal doses (for intragastric administration) have been established: total lethal dose (LD_{100}) 2.5 g/kg; median lethal dose (LD_{50}) 1.045 ± 0.019 g/kg; minimum lethal dose (MLD) 0.1 g/kg.

After a single intragastric administration of calcium boride to rabbits in a dose of 100 mg/kg moderate leukocytosis, a decrease of the content of albumins and increase of the concentration of α-and γ-globulins in the blood serum, fluctuation of the prothrombin index, and increase of the glycotic enzyme aldolase were observed over a period of four months.

Histological investigations revealed slight thickening of the interalveolar septa and in a few cases areas of peribronchial pneumonia. Six months after endotracheal administration of 50 mg of calcium boride an intensified histiocytic reaction was noted in the interalveolar septa, and feeble sclerotic changes after nine months.

The most appreciable pathological changes following the intragastric administration of calcium boride occurred in the liver. Degenerative changes developed in the cells, and nonuniform distribution of glucogen, deterioration of the hepatic trabeculae, accumulation of acid mycopolysaccharides, and sclerotic phenomena in the region of the triad [126] were noted.

Molybdenum boride Mo_2B_5 effected changes similar to those described. Its intraperitoneal injection killed the animals. It should be noted that LD_{50} and the maximum tolerance dose of this preparation are slightly higher than for calcium boride (Table 19), whereas LD_{100} is slightly smaller, being 2.0 g/kg as opposed to 2.5 g/kg for calcium boride.

In chronic experiments on animals slight anemia was noted, as was an increase of the content of amino acid nitrogen in the blood serum, which may have been caused by affection of the enzymic systems catalyzing the conversion of amino acids in the liver or by proteolytic processes in the liver parenchyma, particularly by proteolysis of cell proteins. This is indicated by the degenerative changes in the liver cells.

TABLE 19. Lethal Doses of Borides of Some Metals

Borides	Chemical formula	Lethal dose, g/kg			
		LD_{100}	LD_{50}	MLD	Maximum tolerance
Calcium	CaB_6	2.5	1.045	0.1	0.05
Molybdenum	Mo_2B_5	2.0	1.38	—	1.0

Note. The borides TiB_2, NbB_2, and CrB_2 did not have an acute toxic effect.

Degenerative phenomena developed also in the kidneys, and manifested themselves in swelling of the epithelium of the convoluted tubules.

The toxicity of the other borides was less pronounced. High doses (10 g/kg) of titanium, niobium, and chromium borides failed to kill the animals. The most marked were pneumoconiotic changes in the lung tissue and chronic intoxication phenomena of varying degrees of intensity.

The chronic toxic effect of the compounds is indicated under examination by a loss of body weight effected by zirconium boride [234] and a decrease in the content of hemoglobin and albumins in the blood serum together with an increase of the quantity of the α_1- and α_2-globulin fractions under the action of chromium boride [282].

Bearing in mind the general toxic and fibrogenic effect of the dust of borides, their maximum allowable concentration in the air of industrial rooms in our opinion should be less than the MAC established for metals and their oxides.

On the basis of the data concerning the weakening of the toxic effect of the dust of metal borides on the organism, we can arrange them in the following order: CaB_6, Mo_2B_5, Cr_3B_2, TiB_2, ZrB_2, and NbB_2

The data on the effect of transition metal carbides on the organism [233, 234, 282] and the results of our experiments indicate a lower biological activity of these compounds in comparison with the activity of borides.

On the basis of the data on the greater chemical activity of crystalline boron in comparison with its activity in an amorphous state and of the information on the stronger pathological effect of boron in a crystalline modification, for which a MAC of 5-7 mg/m^3 is recommended, we would recommend 10 mg/m^3 as the MAC for amorphous boron, which has only a marked pneumosclerotic effect.

Comparative Hygienic Evaluation of the Toxic Effect of Transition Metal Silicides

On the basis of the data on the structure of silicides presented in Chapter I and on certain physicochemical characteristics, we can assume that these compounds can differ in their biological activity.

In connection with the industrial use of silicides, particularly in powder metallurgy production, and the possibility of their uptake by workers, we carried out investigations to determine the toxic effect of the main representatives of this group of compounds, namely, $TiSi_2$, $MoSi_2$, and WSi_2.

Fibrogenic and General Toxic Effect of Molybdenum Disilicide $MoSi_2$. Experiments were carried out on 70 white rats to determine the effect of molybdenum disilicide on the organism.

One group of animals was administered endotracheally powdered molybdenum disilicide $MoSi_2$ containing 98% of particles up to 2 μ in size in the form of a suspension in sterile physiological solution. The animals of the second group were subjected to inhalation of molybdenum disilicide in concentrations of 150-250 mg/m^3 (average 215) daily for 4 h for three months and 1-10 mg/m^3 (average 7.7) for six months. The condition and body weight of the experimental animals during the entire observation time were almost the same as for the control animals.

The results of the investigations indicate that molybdenum disilicide has moderate fibrogenic activity. A month after endotracheal administration of the dust the collagen content in the lung tissue increased 18.1% in comparison with the control and continued to increase (Table 20). By six months the increment of collagen was 50.2% ($P < 0.05$).

In the animals that had inhaled airborne molybdenum disilicide for three months at an average concentration of 215 mg/m^3 the collagen level had increased by 42.7% ($P < 0.05$) by the end of the experiment. In the case of the small concentration (average

TABLE 20. Change of Collagen Content in the Lung Tissue at Various Times after the Administration of $MoSi_2$ ($P < 0.05$)

Substance administered	1 month			3 months			6 months		
	Content, mg	Increment		Content, mg	Increment		Content, mg	Increment	
		mg	%		mg	%		mg	%
Molybdenum disilicide	24.7±7.4	3.8	18.1	27.7±5.3	3.3	13.9	28.4±1.1	9.5	50.2
Physiological solution (control)	20.9±1.2	—	—	24.4±3.3	—	—	18.9±4.7	—	—

7.7 mg/m^3) no increment of collagen in the lung tissue was observed during the six months. The morphological changes in the internal organs indicated a general toxic effect of molybdenum disilicide. A month after the administration of the dust phenomena of peribronchitis and perivasculitis, bronchopneumonia, and hyperplasia of the lymph nodes were noted.

In many areas of the lung tissue we found dust particles and dust nodules of small and medium sizes and round shape consisting of lymphoid-histiocytic elements and dust particles.

Three months after the start of the experiment the changes in the lung tissue were analogous to those described; however, the content of dust particles decreased and the lymphoid-cell nodules were larger (Fig. 25).

The protoplasm of the liver cells was swollen with signs of granular degeneration. More pronounced degenerative disorders were observed in the kidneys.

The most marked changes were observed six months after the start of the experiment. Large nodules of lymphoid, histocytic, and fibroblastic elements with a small number of dust particles, moderate diffuse sclerosis, and focal emphysema were detected (Fig. 26).

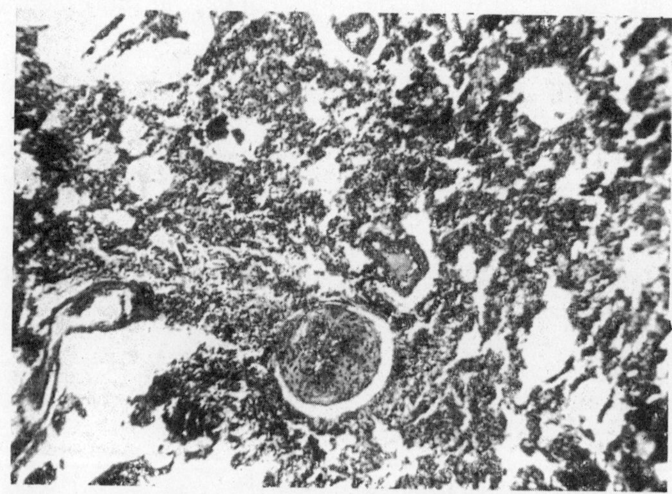

Fig. 25. Productive process of interstition of lung three months after administration of molybdenum disilicide. Magnification 8 × 10.

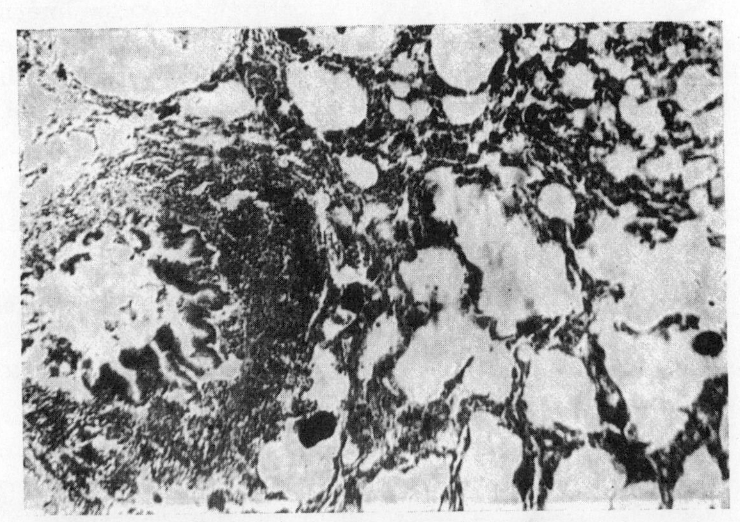

Fig. 26. Pulmonary emphysema and peribronchial infiltration six months after endotracheal administration of molybdenum disilicide. Picrofuchsin staining. Magnification 7 × 10.

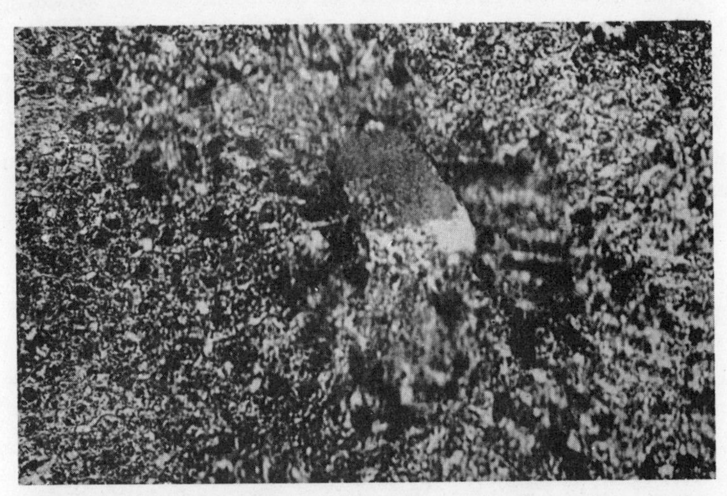

Fig. 27. Fatty degeneration of liver six months after endotracheal administration of molybdenum disilicide. Hematoxylin-eosin staining. Magnification 7 × 20.

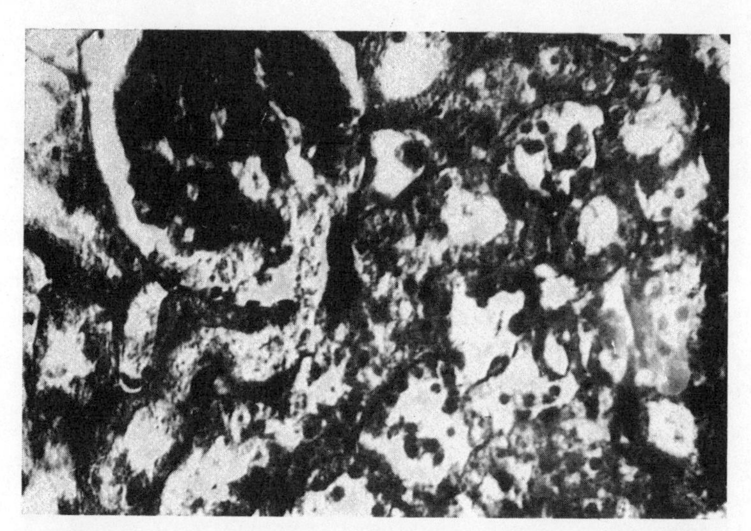

Fig. 28. Marked degeneration of epithelium of renal convoluted tubules with necrotic areas, albuminous exudate in glomerular capsules three months after action of molybdenum disilicide in a concentration of 150-250 mg/m^3. Magnification 7 × 40.

Pathological disorders occurred also in the parenchymatous organs (liver and kidneys). Focal swelling of the liver cells was noted during the first months after the administration of molybdenum disilicide, swelling and granulation of the protoplasm of the cells were observed over more extensive areas after three months, and foci of fatty degeneration were found together with the phenomena described in some animals after six months (Fig. 27).

The morphological changes in the kidneys were characterized by swelling and parenchymatous degeneration of the epithelium of the convoluted tubules. In many glomerular capsules we found an albuminous exudate which apparently was a consequence of an increase of the permeability of the walls of the capillaries of the glomeruli. Similar changes were found in the lungs and other organs exposed to an aerosol of molybdenum disilicide in a concentration of 150-250 mg/m^3 (Fig. 28).

The inhalation of molybdenum disilicide dust at an average concentration of 7.7 mg/m^3 for six months was accompanied by slight changes in the internal organs. In the lung tissue we established small perivascular and peribronchial infiltrates and slight thickening of the vascular walls and interalveolar septa in individual areas of the lungs. There were virtually no changes in other organs. Only in individual animals did we observe, six months after administration of the dust, an accumulation of albuminous exudate in single glomerular capsules and feeble degenerative changes of the epithelium of the renal convolute tubules (Fig. 29).

The results of these investigations show that molybdenum disilicide apparently has a moderate fibrogenic action. However, its toxic effect is not limited to pulmonary lesions. Molybdenum disilicide causes pathological changes in the liver and kidneys. Thus we can conclude that this dust can have a general toxic effect.

Molybdenum disilicide is apparently excreted from the organism gradually, which is indicated by the decrease of the quantity of dust in the lung tissue three months after endotracheal administration and its almost complete absence after 6 months.

The effect of ammonium molybdate on the organism was investigated to obtain some idea of the comparative effect of molybdenum disilicide and other molybdenum compounds, for which a maximum allowable concentration has already been adopted.

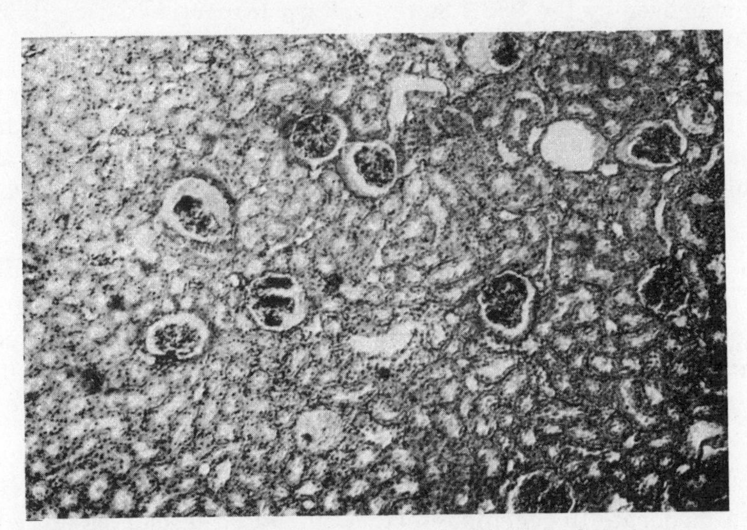

Fig. 29. Parenchymatous degeneration of epithelium of renal convoluted tubules and albuminous exudate in glomerular capsules six months after inhalation of molybdenum disilicide in concentration of 1-10 mg/m^3. Hematoxylin-eosin staining. Magnification 7 × 8.

The morphological alterations in animals induced by ammonium molybdate are similar in their character to the pathological phenomena related with the toxic effect of molybdenum disilicide, but are much more marked. This is apparently due to the high solubility of ammonium molybdate.

Our data are consistent with the results of a study by Mogilevskaya [224], who investigated the toxicity of ammonium paramolybdate, molybdenum trioxide, metallic molybdenum, and molybdenum dioxide, and came to the conclusion that the degree of toxicity of molybdenum compounds depends on their solubility.

Comparing the changes in animals induced by molybdenum disilicide and other molybdenum compounds, we can conclude that molybdenum disilicide is less toxic than molybdenum trioxide and soluble compounds, and is close to metallic molybdenum in toxicity. However, in the case of daily 4-h inhalation of molybdenum in a concentration of 1-10 mg/m^3 (average 7.7) by animals for six months insignificant changes, mainly as a feeble nonspecific reaction in the lung tissue, occurred in the animals. On this basis, we can propose for molybdenum disilicide a maximum allowable concentration in the air of industrial rooms equal to 4 mg/m^3. This MAC was approved by the USSR State Sanitary Inspection Office (USSR SSIO).

Analogous changes in the lung tissue induced by molybdenum disilicide were obtained by Mogilevskaya [230, 233], who, on the basis of results of investigations with endotracheal administration of this compound, established moderate pneumoconiotic changes.

Effect of Tungsten Silicide WSi_2 on an Organism. The effect of tungsten silicide on an organism was studied on 30 white rats. The powdered materials, containing 94.5% of dust particles up to 2 μ in size, was administered to the experimental animals endotracheally as a suspension in sterile physiological solution. The observations were carried out for one, three, and six months.

The results of the investigations indicate that tungsten silicide had a moderate fibrogenic effect. One and three months after the administration of powder the increments of collagen in the lung tissue of the experimental animals were 44% and 51%, respectively. The content of ascorbic acid had also increased by 50.5-56.0%

TABLE 21. Change of the Content of Collagen and Ascorbic Acid (AA) in Lung Tissue after Endotracheal Administration of Tungsten Silicide

Substance administered	Exptl. period, months	Collagen content, mg	Increment of collagen		AA content, mg%	Increment of AA	
			mg	%		mg/100 ml	%
Tungsten silicide	1	30.8±5.3	10.8	51,6	29.8±4.0	10.0	50,5
	3	35.2±6.2	10.8	44.2	28.4±1.5	10.2	56.0
	6	29.7±1.0	10.8	56.9*	17,58±3.46	0.32	1.17
Physiological solution (control)	1	20.9±1.2	—	—	19.8±2.3	—	—
	3	24.4±3.3	—	—	18.2±3,1	—	—
	6	18.9±4.67	—	—	17,9±1.65	—	—

*$P > 0.05$; in other cases $P < 0.05$.

(Table 21). By the sixth month the increase of collagen had reached 56.9%.

A month after the administration of tungsten silicide a histological investigation revealed in the lung tissue thickening of the interalveolar septa due to proliferation of lymphocytes, histiocytes, and other elements, and also dust particles around which were small nodules of lymphoid cells. After three months the changes

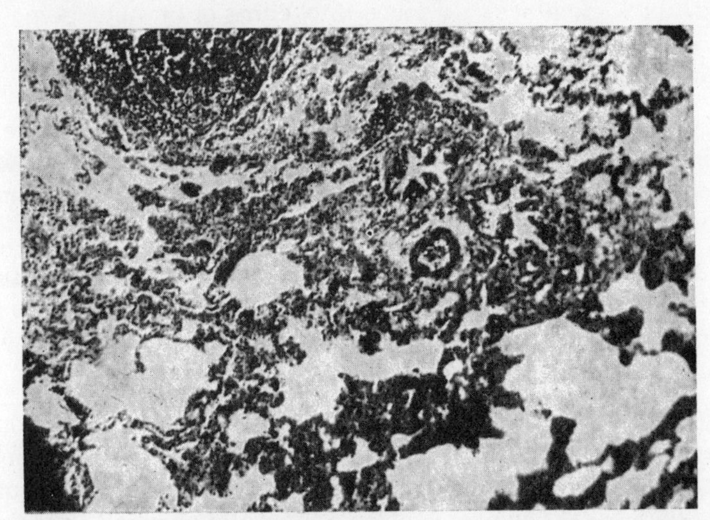

Fig. 30. Cellular-dust nodules in lung tissue, diffuse thickening of interalveolar septa around vessels and bronchi six months after endotracheal administration of tungsten silicide. Magnification 7 × 10.

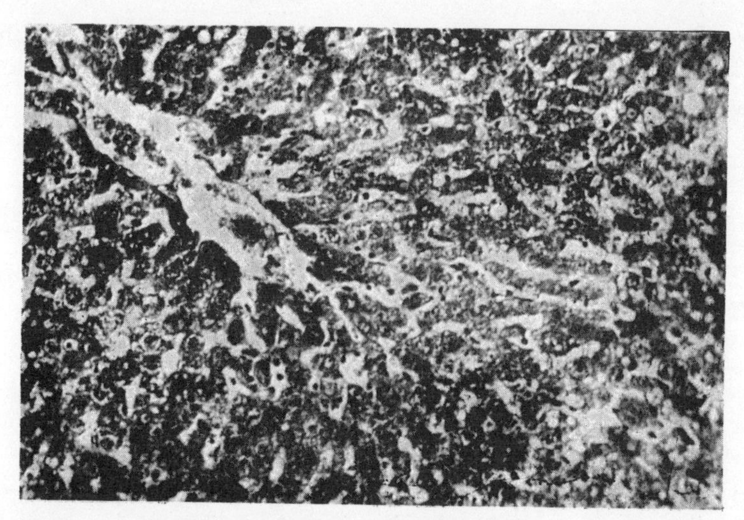

Fig. 31. Degeneration of liver six months after endotracheal administration of tungsten silicide. Magnification 7 × 20.

were about the same as in the preceding case, except there were far fewer dust particles. After six months of observation infiltration by lymphoid elements was noted around the pulmonary vessels and bronchi. The peribroncial lymph nodes were markedly enlarged. In some places we found thickening of the interalveolar septa and aggregates of nodules consisting of lymphoid-histiocytic elements and fibroblasts (Fig. 30).

Characteristic changes occurred in the liver, where we noted swelling of the liver cells and later parenchymatous and sometimes fatty degeneration (Fig. 31). Degenerative changes were found in the epithelium of the renal convoluted tubules.

If we compare the results obtained with the results observed after the endotracheal administration of tungstic oxide (experimental period of 11 months), we arrive at the conclusion that the changes induced are similar.

Tungstic oxide, as N. V. Mezentseva demonstrated, induced thickening of the interalveolar septa, and cellular-dust nodules of lymphoid elements developed. Round-cell infiltrates were noted around the vessels and bronchi, and the walls of small vessels were thickened.

An investigation of preparations of parenchymatous organs showed that the administration of tungstic oxide leads to focal fatty degeneration of the liver and to feeble degenerative changes of the epithelium of the renal convoluted tubules. Analogous changes are observed in animals after the administration of tungsten dust.

Thus, the changes induced by tungsten silicide are similar to the disorders effected by tungstic oxide and metallic tungsten. In connection with this, we must recommend as the maximum allowable concentration for such dust the same MAC as for the dust of tungsten and its oxides, i.e., 6 mg/m^3. The MAC proposed was approved by the USSR SSIO.

Similar data on the morphological alterations in the lung tissue induced by tungsten silicide are given in [205]. On the basis of comparing these changes with the toxic effect of other tungsten compounds, 6 mg/m^3 is recommended as the MAC.

Effect of Titanium Silicide $TiSi_2$ on the Organism. To obtain data on the pathological activity of the dust of titanium silicide $TiSi_2$ its powder was administered endotracheally to 30 white rats as a suspension in sterile physiological solution in an amount of 50 mg per animal. The particle size analysis of the powder included 90% of dust particles measuring not more than 2 μ. Observations were carried out for one, three, and six months.

To compare the toxicity of titanium silicide and its dioxide, a second group of animals was administered endotracheally finely divided powder of titanium dioxide TiO_2 (50 mg), whose effect on the organism had been studied earlier [202, 220, 221, 493, 494].

As early as after a month titanium silicide had induced an increase of the collagen content by 151.6% and after three months

TABLE 22. Change of the Content of Collagen and Ascorbic Acid (AA) in Lung Tissue after Endotracheal Administration of Titanium Silicide

Substance administered	Exptl. period, months	Collagen			AA in lung tissue		
		Content, mg	Increment		Content, mg/100 ml	Increment	
			mg	%		mg/100 ml	%
Titanium silicide	1	52.5±12.7	31.6	151.6	25.8±2.7	6.0	30.3
	3	52.9±2.0	28.5	116.8	14.5±1.5	—3.7 *	—
	6	27.3±4.3	8.4	44.4	17.9±2.26	0.02*	—
Physiological solution (control)	1	20.9±1.2	—	—	19.8±2.3	—	—
	3	24.4±3.38	—	—	18.2±3.1	—	—
	6	18.9±4.67	—	—	17.9±4.65	—	—

*$P > 0.05$; in other cases $P < 0.05$.

by 116.8% ($P < 0.05$). After six months the increment of collagen decreased slightly and was 44.4% ($P < 0.05$); the ascorbic acid level (Table 22) after a month was 30.3% higher ($P < 0.05$).

On investigating histological preparations after three and six months we observed in some areas of the lung tissue a nonspecific inflammatory-proliferative reaction, mainly at sites of dust localization. Cellular-dust nodules formed at later periods, and an interstitial fibrotic process also developed.

Of the other organs the liver and kidneys were affected after the administration of titanium silicide. After a month we observed fatty degeneration in the liver, and after six months this covered extensive areas. In addition, foci of round-cell infiltration appeared around the vessels and in the parenchyma of the liver.

In the kidneys the main symptom was swelling of the epithelial cells of the convoluted tubules, and only in a few animals was an accumulation of exudate noted in the cavity of the glomerular capsules after three and six months.

Similar changes were observed three and six months after the administration of titanium dioxide. A histological investigation after eight months revealed in the lungs phenomena of interstitial pneumonia, thickening of the interalveolar septa, moderate sclerosis, and emphysematous foci. Thickening of the vascular walls and bronchi was accompanied by lymphoid-histiocytic infiltration and enlargement of the peribronchial lymph nodes. Cellular-dust nodules were encountered mainly around the dust aggregates.

Analogous changes in the lung tissue after endotracheal administration of titanium and its dioxide were described by Mogilevskaya and others.

Eleven months after two endotracheal administrations (50 mg each) of the dust of metallic titanium we noted in the lungs aggregates of dust particles around which arose a cellular reaction of a histiocytic and lymphoid character. Round-cell and histiocytic infiltrates with sclerotic phenomena appeared around the bronchi. Hyperplasia and sclerosis of the lymph follicles also developed, and in some areas focal thickening of the alveolar septa appeared.

Eleven months after the administration of titanium dioxide a small amount of dust was noted in the lung tissue, with the formation of round-cell and histiocytic infiltrates at sites of its localization; thickening of the interalveolar septa, small emphysematous areas, and hyperplasia of the lymph follicles around the bronchi and their sclerosis were found.

In view of the fact that titanium silicide produced a more pronounced fibrogenic reaction than did its dioxide, the MAC for $TiSi_2$ should apparently be reduced to 4 mg/m^3, i.e., should be the same as for dust with a high silicon dioxide content. The proposed MAC was approved by the commission of the USSR Ministry of Health.

Phagocytic Activity of Molybdenum and Tungsten Silicide Dust. The effect of various dusts on the organism depends to a considerable extent on the activity of the phagocytic reaction, which determines the rate of elimination of dust particles.

To obtain a better understanding of the effect of silicide dusts on the organism we investigated the phagocytic activity of neutrophils and macrophages after the administration of the powders of molybdenum and tungsten silicides. The possibility of the elimination of these substances from the organism is indicated by a decrease of the content of dust particles in the lung tissue after their endotracheal administration with increase of the experimental period.

Some data on phagocytic activity induced by tungsten powder can be found in the literature [288]. Investigations of the phagocytic activity after the administration of the dust of tungstic oxide and tungsten ores in comparison with the phagocytic activity induced by quartz and coal dust established that the phagocytic reaction of the organism varies depending on the content of quartz. A high content of silica and tungstic oxide suppresses it. The phagocytic activity evoked by these dusts is more similar in its character to the phagocytic activity induced by quartz dust than that due to coal dust. This is indicated by comparatively feeble phagocytosis, the presence of destroyed cells in the exudate, and cytological composition similar to that observed after the administration of quartz dust.

The feeble phagocytic reaction induced by the dust of pure tungstic oxide indicates the suppressibility of the latter with respect to the phagocytes of the exudate. Experimental investigations [3] proved that an increase of intoxication is accompanied by a decrease of the ability of the organism to produce macrophages, as a consequence of which the activity of protective forces drops.

An investigation of the phagocytic reaction to the administration of silicides was carried out on mice of both sexes weighing from 18 to 24 g (by the usual method).

After the administration of molybdenum and tungsten silicides the maximum phagocytic activity was noted 6 h after the administration of the dust, when 50% of the cells participated in the phagocytic reaction.

Under the influence of molybdenum disilicide the intensity of phagocytosis reached the initial level as early as after 48 h, whereas after the administration of tungsten silicide the phagocytic activity reached the initial level only 72 h after administration of the dust. During the first 6 h after the administration of molybdenum disilicide only 30% of the neutrophils were active and later, after 24 h, the activity of the neutrophils dropped to the original values and phagocytosis was accomplished mainly by macrophages.

The phagocytic reaction to the administration of molybdenum disilicide dust is considerably weaker than to the administration of graphite and iron dust. Whereas 3 h after the administration of graphite the intensity of phagocytosis reached 50%, was a maximum (70-90%) by 48 h [288], and reverted to its initial level by 72 h, the phagocytic reaction induced by molybdenum disilicide at the same periods was considerably lower.

The phagocytic activity of neutrophils after the administration of tungsten silicide was lower than that arising after the administration of graphite and iron dust. The phagocytic activity of the neutrophils reached a maximum (27-34%) 24 and 48 h after the administration of the dust and the initial level by 72 h; 48 h after the administration of graphite dust the intensity of phagocytosis reached 70-95% and did not return to the initial level by 72 h. During the first 6 h phagocytosis was accomplished largely by macrophages and thereafter with considerable participation of neutrophils. Our data on the intensity of phagocytosis after the administration of tungsten silicide agree with the data on phagocytosis after the effect of tungstic oxide.

Generalizing the data on the toxic effect of titanium, molybdenum, and tungsten silicides on the organism, we can conclude that the compounds have a fibrogenic action and produce a toxic effect when taken up into the lungs, which manifests itself in the development of nonspecific interstitial bronchopneumonia with cellular-dust nodules and moderate diffuse sclerosis, development of perivascular and peribronchial infiltrates, and perivascular edema (molybdenum disilicide). In addition, all three compounds induce various degenerative changes of the liver (parenchymatous and sometimes fatty degeneration) and of the renal convoluted tubules. Molybdenum disilicide effected an increase of the permeability of the capillaries of the renal glomeruli.

Effect of Refractory Nitrides on the Organism

There are almost no literature data on the pathogenic activity of refractory compounds of the nitride group. One report on the effect of silicon nitride on the organism has appeared recently [7].

The possibility of wide use of these materials in industry and data on the untoward effect of other refractory compounds on workers necessitated an investigation of the toxic properties of nitrides and determination of their allowable concentration in the air of industrial rooms. In this connection, we investigated, in experiments on white rats, the effect of the nitrides of boron, silicon, titanium, zirconium, aluminum, and niobium and boron carbonitride in comparison with the effect of the main components of these compounds [43].

All powders, containing 80-90% of particles up to 2 μ in size, were administered endotracheally by the usual method. In addition, some animals were subjected to inhalation poisoning by the dust of boron nitride and silicon nitride in Latushkina's chambers daily for 4 h in concentrations of 150-200 and 1-10 mg/m^3. The animals were observed for one, three, and six months after the start of the experiment.

Effect of Boron Nitride and Carbonitride on the Organism. Boron nitride BN is a white, fine powder. As investigations showed, its action on animals did not substantially affect the general state of health and body weight, but did cause moderate collagen formation in the lung tissue, i.e., an increase of fibrogenic activity. Three months after the administration of the material the increment of collagen in the lungs of the experimental animals amounted to 50.2% and the content of ascorbic acid had incrased by 19.7% in comparison with the control (Table 23).

The morphological changes in the lungs of all animals were the same and characterized by thickening of the vascular walls

TABLE 23. Content of Collagen in Lung Tissue of White Rats after Endotracheal Administration of Nitrides of Nonmetals

Substance administered	1 month			3 months			6 months		
	Content, mg	Increment		Content, mg	Increment		Content, mg	Increment	
		mg	%		mg	%		mg	%
Boron nitride BN	29.2± ±1.5	—	—	37.7 ±1.3	12.6	50.2	31.7 ±2.4	8.6*	37.2
Boron carbonitride BNC	29.7± ±2.3	0.5	1.7	31.0 ±2.7	9.5	23.5	—	—	—
Silicon nitride Si_3N_4	31.0± ±2.7	1.8	6.1	36.8 ±1.8	11.7	46.6	30.7 ±1.8	7.6	32.8
Physiological solution (control)	29.2± ±2.5	—	—	25.1 ±2.7	—	—	23.1 ±0.3	—	—

*$P > 0.05$; in other cases $P < 0.05$.

and presence of peribronchial and perivascular infiltrates from lymphoid-cell elements. The mucosa of the bronchi was desquamated in place. Mucus with an admixture of desquamated epithelial cells and dust particles were found in the bronchial lumina.

Infiltration from lymphoid elements was observed in the walls of some bronchi. The interalveolar septa at sites of dust aggregates were thickened owing to proliferation of cells of the lymphoid and histiocytic series. Emphysematous foci were also encountered. The cellular-dust nodules consisted of dust particles, dust cells, shadow cells, and large cells with light protoplasm containing dust-particle inclusions but no nucleus.

At later periods (after six months) we noted a decrease of dust particles and dust cells and cells of the epitheloid and lympho-histiocytic series predominated in the nodules; fibrous structures of connective tissue also appeared. Moderate diffuse sclerosis developed in the interalveolar septa; phenomena of bronchitis and peribronchitis with desquamation of the mucous membrane were noted.

In the case of inhalation poisoning of the animals with boron nitride in a concentration of 100-200 mg/m^3 (average 167), too, no changes whatsoever were observed in the behavior of the animals in a period of three months.

After daily 4-h inhalation of boron nitride in a concentration of 100-200 mg/m^3 the collagen content in the lung tissue of these

TABLE 24. Collagen Content in Lung Tissue of White Rats after Inhalation of Nitrides of Nonmetals

Substance administered, mg/m^3	1 month			3 months			6 months		
	Content, mg	Increment		Content, mg	Increment		Content, mg	Increment	
		mg	%		mg	%		mg	%
Boron nitride									
100-200	31.8±1.5	10,3	47.8	30.6±1.8	9,8	47,1	53.4±1,7	30.3	131,1
1-10	31.0±2.8	9.5	44.1	31.9±1.8	11.1	53.3	31,6±3,6	8.5	36.7
Silicon nitride									
100-200	32.2±0.99	10.7	49.4	35.1±1.8	8,3	30.9	42,5±2.3	13.3	57.5
1-10	32.4±0.05	11.2	52.0	36,1±4.08	9.3	34.6	37,4±5.4	8.5	35.5
Physiological solution (control)	21.5±2.1	—	—	26.8±2.8	—	—	23.1±0.9	—	—

Note. For the experimental data $P < 0.05$.

animals after a month increased 47.8% and after six months 131.1% (Table 24). The ascorbic acid level in the lung tissue after three and six months was also raised by 25.8-83.3% ($P < 0.05$).

The morphological changes in the lung tissue were characterized only by an inflammatory-proliferative reaction on the part of the lungs and bronchi. The bronchial lumina contained mucus with an admixture of dust particles and desquamated epithelial cells. Dust particles were found in the lumina of the alveoli and peribronchial lymph nodes. Moderate diffuse sclerosis developed at these sites and cellular-dust nodules appeared.

No significant changes were detected in the animals following inhalation of boron nitride in a concentration of 1-10 mg/m^3 (average 6). Similar data were obtained in the case of endotracheal administration of boron carbonitride BNC to the animals. In this case the collagen content increased by only 23.5%. We observed some swelling of the walls of small vessels and phenomena of peribronchitis and perivasculitis.

In the lung tissue of individual animals we detected small cellular-dust nodules and numerous dust particles localized in the alveoli, interalveolar septa, and peribronchial lymph nodes. Focal thickening of the interalveolar septa and extenstive areas of emphysema appeared, as did feeble diffuse sclerosis at later periods. No pathological changes were observed in other organs.

The changes induced by boron nitride and carbonitride correspond in their character to those which are effected by amorphous boron. However, boron and boron nitride induce considerable changes in the bronchi and irritation of the lung tissue. During the formation of boron carbonitride the boron is apparently bound more strongly as a consequence of reaction not only with nitrogen but also with carbon.

Thus the reaction of the organism to the administration of the aforementioned powdered materials corresponds mainly to that which is observed under the action of dust with a feeble fibrogenic action. Taking this into account, as well as the similarity of the biological effect of the dust of boron and its nitride compounds and that at a concentration of boron nitride of 1-10 mg/m^3 no significant changes were observed in the organism, we can recommend as the MAC for boron nitride dust 6 mg/m^3 and for boron carbonitride 10 mg/m^3. The proposed MAC's were approved by the USSR SSIO.

Toxic Effect of Silicon Nitride Si_3N_4. Silicon nitride is a chemical compound of elemental silicon with nitrogen. It is a light-weight, light-colored powder. Several nitride phases exist: Si_3N_4, Si_2N_3, and SiN.

Up to now the silicon nitride whose chemical composition corresponds to the formula Si_3N_4 has been studied most.

After endotracheal administration of this silicon nitride the condition of the experimental animals during the entire experimental period did not differ from that of the control animals; all animals gained weight.

The collagen content in the lung tissue increased 46.6-32.8% in comparison with the control (see Table 23). The ascorbic acid content in the lungs did not change significantly.

A microscopic study of histological preparations revealed in the lung tissue thickening of the walls of small- and medium-caliber vessels and aggregation of dust particles in the alveolar lumina, around which cells of the epitheloid and histiocytic series formed nodules.

On Van Gieson stained sections we observed slight proliferation of connective tissue in the interalveolar septa, around the bronchi, and in the peribronchial lymph nodes. Morphological changes were absent in other organs.

In the case of inhalation of silicon nitride in a concentration of 100-200 mg/m^3 (average 165) the collagen content increased by 30.9-57.5% ($P < 0.05$) three and six months after daily inhalation (see Table 24).

Morphological changes in the internal organs of this group of animals were negligible: The interalveolar septa, alveoli, vessels, and bronchi did not experience appreciable changes. Occasionally single, small cellular-dust foci were found in the lung tissue.

Silicon nitride in a concentration of 1-10 mg/m^3 (average 5.9) induced only a slight increase of the content of collagen and ascorbic acid in the lung tissue.

The results of investigations of the biological activity of elemental silicon indicate that it is a relative inert substance.

The changes effected by silicon nitride correspond to a considerable extent to those induced by elemental silicon. Taking into account the aforesaid and also that silicon nitride in a concentration of 1-10 mg/m^3 (average 5.9) does not cause substantial pathological phenomena in experimental animals, we recommended 6 mg/m^3 as the MAC for the dust of silicon nitride and silicon in the air of industrial rooms. The MAC for silicon nitride was approved by the USSR SSIO.

Analogous pathological changes in the lung tissue induced by the same substances are described in [7]. Comparing the morphological changes in the lung tissue, the author concluded that the most active is quartz dust and then silicon nitride and elemental silicon.

The report referred to above gives data on the change of the ratio of serum proteins induced by the substances investigated, which indicate stimulation of the reticuloendothelial system. Silicon nitride and silica induced a decrease in the quantity of albumins and of the albumin-globulin ratio and an increase of the level of α-, β-, and γ- globulins.

Changes in the Organism Induced by Metallic Nitrides (TiN, ZrN, AlN, NbN). To determine the effect of transition metal nitrides on the organism, white rats were administered endotracheally titanium nitride TiN, zirconium nitride ZrN, aluminum nitride AlN, and niobium nitride NbN, and the control animals were administered sterile physiological solution.

TABLE 25. Collagen Content in Lung Tissue of White Rats after Endotracheal Administration of Metal Nitrides

Substance administered	1) month			3 months			6 months		
	Content, mg	Increment		Content, mg	Increment		Content, mg	Increment	
		mg	%		mg	%		mg	%
Nitrides									
Titanium	30.0±5.7	0.8*	2.7	32.9±3.2	10.3	45.6	—	—	—
Zirconium	35.8±0.05	6.6	22.7	34.3±0.07	9.2	36.6	35.0±3.9	5.8	29.8
Aluminum	37.8±5.12	8.6	29.4	32.7±2.2	7.6	30.2	37.5±3.9	8.3	21.5
Niobium	35.4±1.2	6.2	21.2	36.0±3.1	10.9	43.4	—	—	—
Control									
Physiological solution	29.2±1.5	—	—	25.1±2.7	—	—	29.2±3.0	—	—

*$P > 0.05$; in other cases $P < 0.05$.

The results of the investigations (Table 25) show that titanium nitride causes an insignificant increase of collagen in the lung tissue (45.6%) in comparison with the control.

The morphological alterations in the lung tissue were characterized by focal thickening of the interalveolar septa and vascular walls, perivascular and peribronchial infiltrations, and phenomena of bronchitis and peribronchitis. Aggregates of dust particles and cellular-dust foci were sporadically noted.

In the liver of the experimental animals we found foci of parenchymatous and sometimes fatty degeneration. Focal swelling of the epithelium of the renal convoluted tubules was noted. These changes resembled the disorders induced by titanium dioxide, but were more pronounced. On the basis of the aforesaid we consider that the MAC for titanium nitride should be less than for titanium and its oxide, namely, 4 mg/m^3. This MAC was approved by the USSR SSIO.

Zirconium nitride also effected a slight increase (36.6–29.8%) of collagen in the lung tissue, which was accompanied by intensified biosynthesis of ascorbic acid. Its content in the lung tissue after a month increased by 16.6% and after three months by 38.3% ($P < 0.02$); after six months the increment of ascorbic acid was 28% in comparison with the control.

The morphological changes in the lung tissue were characterized by thickening of the vascular walls and bronchi, lymphoid-cell infiltration around them, phenomena of bronchitis, presence of cellular-dust nodules, and feeble diffuse sclerosis. Foci of pneumonia occurred in a few areas in the lungs.

In the liver we observed phenomena of fatty degeneration, in the kidneys phenomena of extracapillary nephritis, and also feebly expressed degeneration of the epithelium of the convoluted tubules. In the cortical layer of the adrenal glands we noted a large quantity of lipids, and in the splenic red pulp of some animals a high content of hemosiderocytes.

If we compare these changes with those occurring after the administration of zirconium oxide, we find that the character of the pathological process induced by zirconium nitride corresponds in general to the pathological phenomena effected by zirconium and its compounds. On the basis of this we recommend 4 mg/m^3 (approved by the USSR SSIO) as the MAC for zirconium nitride.

The administration of aluminum nitride was also attended by insignificant changes of the biochemical indices. The morphological changes correspond in the main to those observed under the action of other metal nitrides and consisted in focal thickening of the interalveolar septa; together with this we noted emphysematous areas and thickening of the vascular walls. In the alveolar lumina we encountered conglomerates of dust particles around which had formed an aggregation of lymphoid and histiocytic elements – cellular-dust nodules. Also present were phenomena of bronchitis and at later periods feeble diffuse sclerosis. Some signs of cloudy swelling were observed in the liver. The approved MAC for aluminum nitride is 6 mg/m^3.

Three months after the endotracheal administration of niobium nitride the activity of collagen formation in the lung tissue was about the same as after the administration of other nitrides. The collagen content in the lung tissue at this time had increased by 43.4% and the ascorbic acid content by 5.5% ($P < 0.05$).

The reaction of the lung tissue three months after the administration of niobium nitride consisted in phenomena of bronchopneumonia together with emphysematous foci. Aggregation of lymphoid elements was noted around the dust particles localized in the cavity of the alveoli. The changes induced by niobium nitride in comparison with the effect of niobium oxide and its other compounds are less pronounced [98], and therefore for NbN, just as for niobium, our proposed MAC of 10 mg/m^3 was approved.

Thus the pathological changes induced by metallic nitrides are in general more substantial than those brought about by the action of nonmetallic nitrides (boron nitride and carbonitride and silicon nitride).

Whereas the metallic compounds effect pathological changes in the lungs and parenchymatous organs, the nonmetallic nitrides cause only moderate pneumoconiotic changes.

According to the data of Slutskii [92, 322], the increase of collagen in the lung tissue a month after endotracheal administration of quartz dust was 30 mg, or 195.5%, in comparison with the control and after three months, 90 mg, or 335.3%.

A month after the administration of nitride the increment of collagen was negligible or absent altogether. After the administration of boron nitride the maximum increase of collagen was observed after three months, reaching 50.2%. In other cases the accumulation of collagen was 30-46% and under the action of the most stable compound, BNC, 23.5%.

Consequently, the fibrogenic activity of nitrides is about one-sixth that of quartz dust. The fibrogenic activity decreases even more when organisms are exposed to compounds with a diamond-like structure. Under the action of boron carbonitride the increment of collagen was about one-half that due to the action of boron nitride and one-fourteenth that effected by quartz dust. Some nitrides formed by the reaction of nitrogen with metals, as already mentioned, cause pathological changes in parenchymatous organs in addition to feeble pneumoconiotic reactions.

Toxic Effect of Rare Earth Refractory Compounds

The rare earth elements and their refractory compounds are being used increasingly in technology owing to their attractive physical properties. They are used in the manufacture of devices such as thermal neutron absorbers in nuclear reactors, in radio engineering and electronics as semiconducting materials, etc. [265, 266, 286, 299].

The valuable physicochemical properties of these substances discovered to date are a stimulus for their intensive study and industrial use in the USSR and abroad.

The majority of rare earth elements have been studied little in a toxicological respect, and almost nothing has been reported on the pathogenic activity of their refractory compounds. Among the rare elements the compounds of lanthanides – oxides, chlorides, nitrates, phosphides, etc. – have been studied more. The results of the most thorough investigations [6, 203, 277, 331, 487, 488, 547] indicate that elements of the cerium and yttrium groups have different toxic effects, the yttrium group being more active. The toxicity decreases from the lower lanthanides to the higher, and also depends on the chemical state of the compounds.

Under the action of rare earth elements of the yttrium group in animals, body weight decreased, the hemoglobin level, number of erythrocytes and rod neutrophils, and prothrombin index dropped, and blood clotting time became longer. The oxides of gadolinium and dysprosium reduced the concentration of alkaline phosphatase in the blood.

The rare earth elements of the yttrium group induce pneumoconiotic alterations and degenerative lesions of the parenchymatous organs.

Cellular-dust nodules, sometimes penetrated by collagen fibers, and phenomena of peribronchitis and pervasculitis occurred in the lungs of animals five months after the administration of the oxides and other compounds of ytttrium, lanthanum, dysprosium, gadolinium, and samarium. In some cases the nodules merged to form large conglomerates which alternated with emphysematous areas. Inflammatory-proliferative reaction with thickening of the interalveolar septa and phenomena of bronchopneumonia were noted at sites of dust localization, especially soon after its uptake into the lungs. Moderate diffuse sclerosis, sometimes with necrotic phenomena, was observed at the indicated sites at later periods. Phenomena of granular and fatty degeneration were detected often in the parenchymatous organs. The expressivity of the described changes in the internal organs increased with an increase of the duration of action of the substances.

A slight drop of the hemoglobin level and number of erythrocytes and thrombocytes was also observed in workers exposed to a mixture of elements of the yttrium group. In addition, pharyngitis, rhinitis, laryngitis, dryness and sloughing of the skin of the hands, fissures on the fingers and palms, pruritus of individual areas of the body, pigmentation of the skin, and alopecia were observed [8]. The authors conclude that the toxic properties of elements of the yttrium group are similar to the effect of yttrium itself.

Unlike these rare earth elements, cerium and its compounds do not cause substantial changes in the parenchymatous organs or pulmonary fibrosis. Their only effect is the occurrence of an inflammatory-proliferative reaction in the lung tissue [331, 547].

The oxides of samarium and europium did not result in the death of animals, and morphological changes were found only in the lung tissue in the form of an inflammatory-proliferative reaction and feeble sclerosis at the sites of dust aggregates [204].

Similar data were obtained for the effect of the oxides of gadolinium and dysprosium. In the case of the endotracheal administration of these compounds, morphological changes also occurred only in the lungs and were characterized by the development of dust granulomas (of the foreign body type), a chronic productive inflammatory process in the lungs, and moderate, slowly progressing development of the connective tissue. These changes were more intensive than those induced by neodymium and cerium [231].

As experimental observations showed, a mixture of elements of the cerium group has a more pronounced toxic effect than individual elements. They were found to cause an insignificant increase of the content of erythrocytes, hemoglobin, and band and segmented cells along with a relative decrease of the number of lymphocytes.

A morphological study of the internal organs of animals exposed to a mixture of cerium group elements revealed confluent pneumonia with the formation of abscesses, destructive purulent bronchitis, dust-induced lymphostasis with a histiocytic reaction, pneumosclerosis, fibrinoid pleurisy, nodular hepatitis with a phagocytic reaction, and cloudy swelling of the epithelium of the renal convoluted tubules.

A comparison of the intensity of development of connective tissue in the lungs induced by rare earth elements indicates that the fibrogenic effect diminishes on passing from yttrium oxide to lanthanum and cerium oxides.

It should be emphasized that, in addition to a fibrogenic effect, a decrease of blood coagulability is characteristic of rare earth elements. Some authors include the rare earth elements among blood antiprothrombins and anticoagulants [55, 476]. Considerably slower blood clotting is observed under the action of neodymium, praseodymium, cerium, and other elements. A toxic effect has been established also for a number of rare earth compounds.

The fluorides of some rare earth elements caused pneumonia, which led to the death of experimental animals [7, 27, 281, 547].

The chlorides of rare earth elements (lanthanum, cerium, neodymium, praseodymium) had a mild toxic effect following administration into the lungs and subcutaneously; in the case of ingestion with food the toxicity decreased. In the case of intravenous injection of rare earth chlorides an increase of their toxicity was noted; inflammatory changes in the lung tissue, edema, cirrhosis of the liver, portal stasis, and pulmonary hyperemia occurred. Granulomatous peritonitis with serous or hemorrhagic ascites developed secondarily [268, 561]. In the case of inhalation the chlorides are rapidly removed from the lungs, but their removal from the organism is much slower following entry into the stomach or through subcutaneous injection.

Cerium, like many other metals, accumulates in the liver and bones [52]. Like chlorides, the rare earth nitrates are localized at the sites of uptake. The nitrates of lanthanum, yttrium, and neodymium, unlike their oxides, can be the cause of slowly healing necroses and ulcerative abscesses with epilation.

Unlike the chlorides, the nitrates spread rapidly in the organism, rendering general toxic effect (pulmonary edema, lesion of the kidneys, lethargy, depression).

The acetates of rare earth elements, having high stability, are eliminated from the organism unchanged following ingestion. In the case of intraperitoneal injection they cause necrosis, adhesive peritonitis with ascites, and acute lesion of the lungs and liver, more rarely of the kidneys.

The results of investigations undertaken to determine the biological activity of certain rare earth refractory compounds, carried out by Olefir [250] under our supervision, provide evidence of their toxic effect.

Samarium hexaboride SmB_6 and lanthanum sulfide LaS, as well as the oxides Sm_2O_3 and La_2O_3, caused the death of experimental animals, whereas yttrium hexaboride YB_6, lanthanum hexaboride LaB_6, and cerium sulfide CeS did not result in death.

All experimental animals were distinguished by lagging in weight gain. An increase of the relative weight of the internal organs under the effect of the refractory compounds did not always occur. An increase of the weight coefficients of the internal organs under the action of the oxides of yttrium Y_2O_3, lanthanum La_2O_3, and samarium Sm_2O_3 was more characteristic.

A study of the fibrogenic activity of the materials under consideration established a feeble fibrogenic effect of the oxides and refractory compounds. A statistically significant increase of the collagen level in the lung tissue was observed only under the action of samarium oxide.

Unlike the oxides, the hexaborides did not cause an accumulation of collagen, whereas the administration of sulfides was accompanied by an increase of collagen formation by a factor of 1.5-2.

A study of the leukocytic reaction to the administration of the oxides and refractory compounds revealed that oxides and hexaborides briefly stimulated leukopoiesis, but subsequently, from the 30th day to the end of the period of observations (six months), the level of leukocytes did not exceed the initial level in the majority of animals. In some cases the sulfides (e.g., cerium sulfide) caused slight leukopenia.

It was established that the hexaborides and sulfides of yttrium, lanthanum, samarium, and cerium, like the oxides of these elements, effect a decrease of blood coagulability which persists for 4.5-6 months after a single administration of the compounds. In this case the blood clotting time normalized before the formation of the first thread of fibrin. A decrease of the prothrombin index by a factor of 2-3 was noted already on the 15th day after administration of the oxides and hexaborides of yttrium, lanthanum, and samarium, and in the sixth month the index was still below its initial level for most animals. The exception was the group of animals that had received cerium oxide, for which the prothrombin index did not change. This index also did not change during the first months of the experiment under the action of the sulfides. Only after three months did a decrease of this index become significant.

Consequently, the oxides, hexaborides, and sulfides of a number of rare earth elements manifest anticoagulant properties in the early period of their action.

The high morbidity of workers exposed to toxic substances is determined to a considerable extent by the state of natural resistance of the organism. In connection with this, we determined the indices of natural immunity (content of lysozyme, complement, and β-lysine in the blood serum) after the action of the aforementioned rare earth compounds and their oxides.

The results of the investigation showed that at early periods of an experiment (15 days) hexaborides lower the lysozyme titer in the serum by a factor of 10 on the average and sulfides by a factor of 3-4. Restitution of the complement titer began a month after the administration of the substances, but complete restitution in the animals that had received sulfides occurred much earlier (after three months) than in the animals subjected to the action of hexaborides (after six months).

The level of lysozyme changed simultaneously with the change of the complement content in the serum.

In animals that had received the oxides and sulfides of lanthanum and cerium we determined the titer of β-lysine, which, like the preceding indices, decreased by the 15th day of observation and then slowly increased. However, even after six months of the experiment its content had not reached the initial level.

The data presented indicate a decrease of natural immunity of the organism under the action of a number of refractory rare earth compounds and oxides, which is not restored for a long time. The oxides have the greatest pathogenic activity, and are followed in this respect by the hexaborides, and finally the sulfides.

Yttrium, lanthanum, and samarium hexaborides caused morphological changes predominantly in the lung tissue, which were characterized by inflammatory-proliferative changes of the pulmonary parenchyma, vessels, and bronchi. Focal thickening of the interalveolar septa appeared owing to proliferation of the lymphoid-cells and histiocytic elements. Desquamated epithelial cells and dust particles were found in the lumina of the large bronchi. Three months after the administration of the substances single lymphoid-cell nodules formed, consisting of dust particles, lymphoid elements, histiocytes, and fibroblasts.

In contrast to this, under the action of the rare earth oxides we observed during the same period a multitude of cellular-dust nodules containing connective-tissue fibers. Sometimes racemose dilation of the bronchi with metaplasia and foci of diffuse sclerosis and emphysema occurred. Thus, the morphological alterations effected by hexaborides are somewhat less pronounced than those induced by oxides.

We obtained similar results in a histological investigation of the internal organs of animals poisoned with rare earth sulfides. They also induced inflammatory-proliferative changes in the lung tissue and bronchi, but these were more feeble than those effected by the oxides and borides.

An investigation of the phagocytic reaction of the organism to the administration of powdered oxides, hexaborides, and sulfides of rare earths established a high activity of macrophages.

Yttrium, lanthanum, samarium, and cerium oxides caused an increase of the intensity of phagocytosis as early as in the first hours of the action of the dust. Approximately the same level of phagocytosis was noted after the administration of the hexaborides. Phagocytosis effected by samarium oxide and hexaboride was distinguished by an especially high intensity, whereas the action of cerium oxide was accompanied by suppression of the phagocytic reaction.

During the first hours after the administration of lanthanum and cerium sulfides the intensity of phagocytosis was maximum and considerably higher than after the administration of quartz dust.

Rare earth hexaborides and oxides also caused intensified phagocytosis of dust particles by cells. The phagocytic number, i.e., the average number of dust particles engulfed by one active cell, was at a high level under the action of the hexaborides during the entire period of observation (72 h). This index dropped noticeably by the end of 72 h only under the action of cerium oxide. As a result of the action of the sulfides the phagocytic number approximated that induced by silica dust. In this case we noted considerably fewer phagocytes than after the administration of the oxides and hexaborides.

A study of the morphological composition of the exudate showed that during the first days phagocytosis was accomplished by neutrophils and occasionally by monocytes. There were almost no histiocytes on the preparations. At late periods of the experiment the number of histiocytes increased to 35-40%. After 48-72 h we noted a high content of destroyed and semidestroyed cells, and also freely lying dust particles.

Thus the hexaborides, sulfides, and oxides of a number of rare earths are actively phagocytized and, we must assume, are intensely removed from the organism, which should be regarded as a manifestation of a protective reaction aimed at reducing the effect of the toxic agent.

In addition to the investigations presented, we investigated the effect of oxides of rare earths and refractory compounds on some metabolic processes. The data obtained indicate changes of protein metabolism and oxidation-reduction processes in the organism under the action of the substances indicated. Lanthanum, yttrium, and samarium hexaborides and lanthanum and yttrium oxides effect an insignificant decrease of albumins and the albumin-globulin ratio, and also an increase of the level of globulins, particularly α-globulins. We also noted a decrease of the quantity of SH groups in the blood, serum, and tissues of the internal organs accompanied by a decrease of the general activity of oxidation-reduction enzymes.

A comparison of data on the effect of refractory compounds on the organism with literature data on the toxic effect of oxides permits the conclusion that the character of the pathological changes induced by refractory compounds corresponds in general to the toxic effect of rare earth oxides, but the toxic effect in the former case is less pronounced.

Rare earth compounds can be arranged in the following order of decreasing intensity of toxic manifestations: oxides, hexaborides, and sulfides. The difficulty of producing refractory compounds and their high cost complicate the possibility of conducting investigations with inhalation poisoning of animals for the purpose of establishing allowable concentrations.

Comparing the data presented on the effect of rare earth elements and their compounds on the organism with the data on the effect of other refractory compounds and their oxides on the organism, we can recommend preliminarily as the MAC for oxides and refractory compounds of the yttrium group 4 mg/m^3 and for the cerium group 6 mg/m^3.

Summing up the experimental data presented in this chapter, we must point out that refractory compounds have a pathological effect whose degree of manifestation is related with their chemical composition and crystal structure.

Refractory compounds based on nonmetals are characterized by the development of feeble pneumoconiotic changes and the presence of bronchitis whose degree of manifestation depends on the activity of the components forming the compound. Under the action of refractory compounds obtained by reacting the transition d and f metals with nonmetals we observed, in addition to pneumoconiotic changes and bronchitis, signs of a toxic effect similar to the symptoms of intoxication by the principal metallic components.

The toxicity of refractory compounds is indicated by degenerative changes (cloudy swelling and fatty degeneration) in the parenchymatous organs, primarily in the liver and kidneys, and by changes of other indices. Refractory compounds can be arranged in the following order of decreasing biological activity: borides, silicides, nitrides, and carbides.

The rare earth refractory compounds are less active than the oxides; they are similar in mode of action to the original rare earth elements.

CHAPTER VI

Toxic Effect of Some Chalcogenides

The chalcogens are ore-forming elements (sulfur, selenium, and tellurium) which on reacting with metals, nonmetals, and semimetals (boron, silicon, germanium, tin, and lead) form chalcogenides – sulfides, selenides, and tellurides.

According to the classification proposed in [311], three groups of compounds are distinguished: chalcogenides of s metals; chalcogenides of transition metals with d and f electron shells being filled; and compounds of chalcogens with nonmetals and semimetals having outer p electrons.

The chalcogenides of the alkali metals are characterized by an ionic-covalent bond with the appearance of some contribution of a metallic bond in the tellurides. In connection with this it is suggested [311] that these compounds, and particularly the sulfides and selenides, have semiconducting properties. The chalcogenides of the alkaline earth metals are characterized by the same properties.

The shares of both the metallic bond and the ionic-covalent bond increase in chalcogenides of the copper group.

The chalcogenides of transition metals are characterized by the presence of a covalent bond between the atoms of sulfur, selenium, and tellurium, and also of a metallic and ionic bond between the metal atoms. Thus, the formation is observed of heterodesmic compounds with different proportions of the covalent and metallic bonds, which vary depending on the acceptor capacity of the transition metals. The chalcogenides of lanthanides have bonds of about the same character [115, 299].

Most sulfides have a cubic lattice of the NaCl type, rhombohedral of the $CaCl_2$ type, cubic of the zinc blende type, tetragonal, cubic of the spinel type, and others [311].

The sulfides of alkali and alkaline earth elements, which dissolve in water, becoming hydrolyzed, and also react readily with acids, are the most active chemically.

Sulfides are used most often in technology as semiconducting materials for recording radiations and in photoresistors. The practical use of sulfides is related also with their other properties (conductivity, catalytic activity, refractoriness, lubricating properties, etc.).

Selenium and tellurium also form compounds with almost all elements of the periodic system except inert gases, carbon, nitrogen, and iodine [152, 249, 371, 372].

Selenides and tellurides are characterized by the formation of solid solutions of variable compositions. This pertains especially to tellurides and is apparently related with the stronger metallic properties of tellurium in comparison with selenium. (Thus, the lower stability of tellurides.) On heating such compounds tellurium is split off with the formation of metal, as is the case, for example, with molybdenum and tungsten tellurides.

The selenides of alkali and alkaline earth metals crystallize with the formation of a face-centered cubic lattice of the CaF_2 and NaCl types (for alkaline earth metals) and with a predominance of the ionic bond. Selenides of transition metals form various hexagonal structures with a change of the bond from ionic to covalent.

Tellurides are characterized by a large contribution of covalent bonding as a consequence of the smaller difference of electronegativity between the chalcogen and metal.

Selenides and tellurides have semiconducting properties, and are also used as lubricating (antifriction) additions to mixtures and alloys. From a chemical standpoint, the selenides and tellurides of the alkali and alkaline earth metals are the most unstable. In the presence of moisture they are slowly hydrolyzed. The tellurides obtained as a result of the reaction of tellurium with Group III elements also have low resistance to hydrolysis.

The selenides and tellurides of transition metals are relatively stable chemically, but in humid air they gradually decompose with the liberation of hydrogen selenide and hydrogen telluride.

Toxic Effect of Some Metallic Sulfides

The effect of sulfides on the organism has been described only in single reports concerning the toxicity of the sulfides of alkali metals (potassium and sodium). According to [157], the ingestion of several grams of sodium sulfide was the cause of fatal poisoning as a consequence of the formation of hydrogen sulfide in the gastrointestinal tract. Cases of hydrogen sulfide poisoning have occurred in work in contact with sodium sulfide. Sodium sulfide has a caustic and irritating effect on the skin, mucous membranes of the eyes, and on the air passages [1, 379].

Other reports [131, 379] indicate affection of the fingernails (peeling, loosening, erosion) in workers coming into contact with sodium sulfide [131].

In view of the prospects of wide use of sulfides in industry and our lack of knowledge of these substances in a biological respect, we carried out investigations to determine the toxic properties of some of the most widely used compounds based on transition metals, namely, the sulfides of molybdenum MoS_2 and tungsten SW_2.

The experiments were conducted on 120 white rats which were administered powders of the materials containing 90-96% of particles up to 2μ in size. All the compounds investigated were administered once, endotracheally, in an amount of 50 mg per animal. In addition, poisoning with molybdenum disulfide was accomplished by inhalation in Latushkina's chamber in concentrations of 1-10 (average 6.7) and 150-200 mg/m^3 (average 168.4). The animals poisoned endotracheally were observed for one, three, and six months. The animals were observed for two months in the case of inhalation poisoning at a concentration of 168.4 mg/m^3 and for five months at a concentration of 6.7 mg/m^3.

The results of the investigations showed that molybdenum and tungsten disulfides produce a feeble fibrogenic reaction. An increase of the collagen content by 38.8 and 47.8% was noted only one month after endotracheal administration of these preparations. Thereafter the deviation in the collagen content from its level in the control animals was statistically insignificant. The ascorbic content in the lung tissue did not change substantially either. No regular changes of the quantity of nucleic acids were observed three and six months after endotracheal administration of tungsten disulfide.

The activity of transaminases [255] a month after the administration of molybdenum disulfide was lower by 25% for aspartate aminotransferase and by 48.9% for alanine aminotransferase. After six months we noted a slight (16%) but statistically significant increase of alanine aminotransferase.

The slight increase of the activity of alanine aminotransferase after the administration of tungsten disulfide was insignificant.

Pathomorphological investigations a month after the administration of sulfides revealed only slight changes in the lung tissue. They were characterized by focal thickening of the interalveolar septa due to proliferation of lymphoid-cell and histiocytic elements around the vessels and dust particles. The number of the latter decreased with increase of the experimental period and after six months they were found only in single cases. An enlargement of the peribronchial lymph nodes was noted predominantly in the initial period of observations in some animals, whereas at later periods emphysematous phenomena were noted more often.

After inhalation poisoning of the animals with molybdenum disulfide in concentrations of 1-10 (average 6.7) and 150-200 mg/m^3 (average 168.4) we were not able to find regular and significant deviations in the content of collagen and ascorbic acid or of other indices from the norm.

A study of histological preparations showed that morphological changes, too, had occurred only in the lung tissue and were manifested by a feeble inflammatory-proliferative reaction. Areas of thickening of the interalveolar septa appeared at sites of dust localization. Appreciable morphological changes were absent in animals exposed to molybdenum disulfide in a concentration of 1-10 mg/m^3.

A study of the phagocytic activity under the action of molybdenum and tungsten disulfides established that both dusts were actively phagocytized. The intensity of phagocytosis during the first three hours after the administration of molybdenum disulfide reached 94%, dropped slightly thereafter, and after 72 h amounted to 84% of the control.

After the administration of tungsten disulfide the intensity of phagocytosis was slightly lower, reached 76% after 3 h, and stayed at about the same level during the remaining period of the experiment.

The intensity of the phagocytosis was close to that induced by graphite, which is effectively removed from the organism.

Taking into account the feeble activity of molybdenum and tungsten disulfides, the absence of signs of intoxication in the animals poisoned by these materials, and the good ability of the organism to eliminate them, we consider it possible to regard these sulfides as low-activity dusts. Since after the inhalation of molybdenum disulfide in a concentration of 1-10 mg/m^3 by the animals we were not able to detect any substantial changes in the organism, we can recommend 6.0 mg/m^3 as the MAC for molybdenum disulfide dust.

We can recommend as the MAC for tungsten disulfide the same concentration as for molybdenum disulfide, since they have a similar effect on the organism.

A comparison of the pathogenic activity of both types of dust indicates that molybdenum disulfide is more active. This is confirmed by the increase of collagen and presence of pronounced inflammatory changes in the lung tissue after the administration of molybdenum disulfide to the animals and by the results of a study of phagocytic activity.

A comparison of the literature data and the results of our observations permits establishing that the sulfides of alkali metals, being less stable chemically, are more active biologically than the sulfides of transition metals of high sulfur content, in which covalent bonds predominate.

This apparently explains the low pathogenic activity of rare earth sulfides.

Toxic Effect of Selenides and Tellurides of Transition Metals

There are no data on the effect of selenides and tellurides on the organism. The toxic effect of selenium and tellurium has been described by many authors. Both elements cause acute and chronic poisoning. This is indicated by results of medical examinations of workers and clinical observations of the state of health of persons exposed to high concentrations of these substances in cases of industrial accidents [107, 109, 171, 236, 237, 399, 423].

Comprehensive experimental investigations [102, 173, 355, 557] indicate a higher toxicity of selenium dioxide in comparison with metallic amorphous selenium.

The substances under consideration caused the death of animals and effected pronounced changes in the lungs and bronchi (inflammation, atelectasis), plethora of the thoracic and abdominal organs, intraalveolar edema of the lungs, and degenerative changes of the hepatic and renal tissue. In chronic inhalation poisoning phenomena of bronchopneumonia and peribronchitis occurred; in the liver, hemorrhages, focal necrosis, fatty degeneration; necrotic and degenerative changes of the epithelium of the renal convoluted tubules. Functional and pathomorphological alterations of the reproductive organs were found. The passage of selenium through the placenta lowered the viability of the fetus and caused its death.

A study of the distribution of selenium and hydrogen selenide in the organism revealed maximum selenium concentration in the liver and kidneys, and also in the blood, lungs, and spleen. Its content in the brain, bones, and muscles was negligible. Selenium was eliminated mainly with the urine.

Many authors suggest that underlying the pathological disorders are changes of the erythrocytes which are manifested in the development of hypochromic anemia [557]. In the opinion of other authors, intoxications are related with disturbances of protein metabolism, oxidation-reduction processes, and other processes as a consequence of inactivation of enzymes.

In addition to these disorders, data are given in [102, 113] on a decrease of the blood glutathione level as a result of the direct binding of sulfhydryl groups and conversion of reduced glutathione to oxidized, and also in connection with disturbance of its synthesis.

The concentration of free, mainly sulfur-containing amino acids in the blood dropped under the action of selenium. A change of the content of protein fractions in the blood serum was also noted: A drop of the level of albumins and an increase of the level of β-and γ-globulins. The rate of protein metabolism decreased as a consequence of blocking of the active SH groups of the protein molecules.

On the basis of experimental and industrial observations the following MAC's are recommended for the air in industrial rooms: 2 mg/m^3 for selenium and 0.1 mg/m^3 for selenium dioxide.

Tellurium, in the opinion of the majority of authors [413, 464], has a greater toxic effect than selenium. Tellurium dioxide is characterized by a more pronounced toxic effect than elemental tellurium. Tellurites (salts of tellurous acid H_2TeO_3) and tellurates (salts of telluric acid H_2TeO_4), which are soluble in body fluids, are more toxic than elemental tellurium and its dioxide [287, 443]. According to some authors, tellurium and its compounds cause acute and subacute poisoning [281, 458, 464].

In acute poisoning by hydrogen telluride marked irritation of the mucous membranes of the air passages, which were stained black-green, occurred. It was established that hydrogen telluride has a weaker toxic effect than hydrogen selenide and hydrogen arsenide.

Other tellurium compounds cause headaches, weakness, dizziness, increase of the respiration and pulse rates, dyspnea, nausea, vomiting, garlic odor of the breath and perspiration, renal pains, hematuria, cystitis, and cyanosis. A comatose state occurs before death. Yellow discoloration of the subcutaneous tissue, specific odor, black coloring of the mucosa of the gastrointestinal tract, plethora of the internal organs, degeneration and edema of the liver, and inflammation of the urinary tract were noted at autopsy [464, 474, 562].

In chronic poisoning by tellurium in small concentrations the state of workers was characterized as follows: garlic breath, garlic odor of perspiration and urine, dryness and metallic taste in mouth, nausea, vomiting, insomnia, depressed state, hyphidrosis, digestive disorders (diarrhea or constipation), emaciation. Such states occurred after working two years in an atmosphere containing 0.01-0.05 mg/m^3 Te. Up to 400.06 mg/liter Te were found in the urine of all examinees. A metallic taste in the mouth was perceived at its content of 0.01 mg/liter. There were no such complaints in the case of a lower tellurium content. The characteristic odor of the breath was not perceived when tellurium was absent in the urine; there were no changes in the state of health.

After administration of tellurium dioxide to animals their growth was retarded, their general condition deteriorated, and paralysis of the hindlimbs and oliguria or anuria occurred. Several degenerative changes appeared in the kidneys, spleen, heart, lungs, and liver, and less serious affections in the bones, skeletal muscles, and brain. The expressivity of the pathological alterations in the structure of the internal organs was consistent with the data on the buildup of radioactive tellurium in them [417, 418]. The content of hemoglobin and erythrocytes changed, and other symptoms of the toxic action appeared, particularly disorders of tissue respiration as a result of blocking of SH groups [562].

The appearance of a garlic odor in the breath of white rats (just as in people) was noted after repeated oral administrations of elemental tellurium in doses of 6-19 mg. The garlic odor was detected as early as 24-32 h after the administration of tellurium and ceased only several days after the end of the experiment.

The administration of ascorbic acid promoted the disappearance of the unpleasant odor [418, 569]. This was due to the binding of tellurium by the ascorbic acid molecule and to the prevention of formation of $Te(CH_3)_2$. However, the compound of tellurium with ascorbic acid is unstable, as a consequence of which the odor returned when administration of the vitamin was stopped.

The morphological changes found in the experimental animals after daily administration of TeO_2 were characterized by focal, fatty, and granular degeneration of the liver and reversible degenerative changes of the epithelium of the renal convoluted tubules. Sometimes pneumonia and bronchitis occurred. In some instances confluent necrotic foci in the liver, marked degeneration of the epithelium of the renal convoluted tubules, and swelling and destruction of the cells of the cortical substance of the adrenal glands were observed. No changes were noted in the central and peripheral nervous system of these animals. After long feeding of tellurium dioxide to cats the mucous membrane of the fundus ventriculi was stained black and degenerative changes (dissapearance of the pyloric glands) and pronounced phenomena of hepatitis developed [417, 418, 443].

Poisoning of white rats (2 h at a time for 13-15 weeks) with a condensation aerosol at a concentration of 50 mg/m^3 (the size of 90% of the particles was less than 1 μ) caused lagging in weight, coarsening of the hair, localized loss of hair, inhibition of excitatory processes in the cerebral cortex, and increase of chronaxie with subsequent paresis of the hindlimbs; six of the ten animals died [287]. Hypochromic anemia developed, catalase activity decreased, the relative content of albumins dropped, the quantity of β-globulins increased, as did α^2- and γ-globulins but to a lesser degree, and the content of free SH groups of the serum declined. The synthetic function of the liver was impaired in some cases. Autopsy of the animals revealed an increase of the relative weight of the liver, kidneys, spleen, and lungs, degenerative changes in the liver and proliferation of the stroma, foci of vacuolation in the kidneys, and increased blood filling of the glomeruli.

The deposition of elemental tellurium in macrophages of the alveolar epithelium was noted in the lungs. Experiments with the administration of condensation and disintegration aerosols showed that the latter were much less toxic. This permits recommending a tenfold increase of the maximum allowable concentration for disintegration aerosols in comparison with the MAC adopted for condensation aerosols, which amounts to 0.01 mg/m^3.

Tellurium and its compounds are easily absorbed from the gastrointestinal tract [417, 418]. Having entered the organism, tellurium oxides, on coming into contact with the tissues, are reduced to elemental tellurium or are converted to $Te(CH_3)_2$, which has a characteristic garlic odor. As investigations [287] showed, when tellurium enters the blood 90% of it is contained in erythrocytes, as a result of which catalase activity is inhibited and hemolysis of the erythrocytes occurs. Maximum quantities of tellurium are found in the kidneys and smaller amounts in the liver.

Tellurium is eliminated predominantly through the kidneys and to a lesser degree through the gastrointestinal tract. After gastric administration of tellurium a large part of it is excreted with the urine. Thus data on the effect of tellurium and its compounds on the organism provide evidence of their toxicity. The toxic effect from the administration of tellurium and its compounds is more pronounced than in the case of the administration of selenium and its compounds.

To obtain data on the comparative toxicity of selenides, a group of animals was administered endotracheally metallic selenium and $NbSe_2$, $MoSe_2$, and WSe_2, i.e., selenides of transition metals. In addition, some animals were subjected to inhalation poisoning by tungsten diselenide in Latushkina's chamber in concentrations of 1-10 mg/m^3 (average 3.7) for one, three, and six months and 150-200 mg/m^3 (average 167) for one and three months, daily for 4 h. The general condition and weight of the animals was checked during the experiments.

The compounds investigated contained 63-66% of selenium and 34-37% of metals. Smaller particles predominated in the composition of the tungsten diselenide powder: 44% up to 1 μ and 84% up to 2 μ in size. The particle size of the selenides of niobium, molybdenum, and selenium was also small. Particles up to 2 μ in size amounted to 66-70%.

The results of biochemical investigations were processed statistically, with determination of the mean statistical error and calculation of the percentage degree of confidence of the data obtained (P) by a simplified method [238].

The collagen content in the lung tissue of white rats that had received metallic selenium increased by 33% in comparison with the control (P = 2%) a month after administration; no substantial difference was noted in the collagen content in the lung tissue of the experimental and control animals after three and six months.

A tendency toward an increase of the collagen content in the lung tissue appeared also a month after endotracheal administration of niobium and molybdenum diselenides, whose level in the lung tissue of the experimental animals in comparison with the control had increased by 51.3 and 36.7%, respectively. However, the degree of confidence of these changes is small (P = 10 and 14%). A significant (27%) increase of the collagen content of the experimental animals was observed three months after the administration of molybdenum diselenide, whereas after six months its level in the lung tissue of the experimental animals did not differ from that in the control animals.

After the endotracheal administration of tungsten diselenide the collagen content did not increase and amounted to 74.1 and 70.9% of the control (P = 4 and 0.1%).

A weak fibrogenic activity of tungsten diselenide is indicated by the results of experiments with inhalation poisoning of animals. For a concentration of 150-200 mg/m^3 the collagen content in the lung tissue increased maximally by 23.9% in comparison with the control (P = 7%).

In the lung tissues of animals exposed to tungsten diselenide in a concentration of 1-10 mg/m^3 a significant (25.7%) increase of the collagen content was observed only six months after the administration of the compound.

In animals that had received metallic selenium we noted a more marked decrease of sulfhydryl groups in the blood, serum, and brain and heart tissues; it occurred as early as a month after administration of the preparation and subsequently recovered slowly. The most intense collagen formation occurred at this time. Unlike selenium, the selenides induce collagen formation much later. In some cases it was expressed feebly, as, for example, after the administration of tungsten diselenide. Under the action of selenium the activity of aminotransferases was a maximum in the first month after the start of the experiment, whereas after the administration of tungsten and niobium diselendies only a tendency toward an increase of the activity of aspartate aminotransferase appeared and the activity of alanine aminotransferase alone increased with statistical significance; the increase of activity of both aminotransferases was insignificant after administration of molybdenum diselenide. After three months the activity of the enzymes as a result of the administration of the three preparations dropped, and only after the administration of metallic selenium did aspartate aminotransferase activity remain at a maximum. After six months the activity of both aminotransferases in the animals that had received metallic selenium was close to the activity in the control animals, and after the administration of niobium diselenide and especially molybdenum diselenide was still significantly high. About the same regularity was observed in the case of inhalation poisoning of the animals with tungsten diselenide in a concentration of 1-10 mg/m^3.

Six months after the administration of the selenides the content of γ-globulin was elevated with statistical significance; this being true to a greater degree for tungsten diselenide than for molybdenum diselenide (by 21 and 106%, respectively). However, the most pronounced deviations on the part of the serum protein fractions were induced by metallic selenium.

A determination of the content of nucleic acids showed that the changes induced by selenium and selenides pertain more to DNA than to RNA; the DNA content in the liver is especially diminished, which was noted as early as a month after the administration of the most toxic materials – metallic selenium and tungsten diselenide. Under the action of the latter the quantity of DNA was statistically lower also six months after the start of the experiment. Under the action of less toxic substances (for example, after the administration of molybdenum disulfide), the decrease of the DNA content in the liver after a month was statistically insignificant, and became significant only after six months; three months after the administration of niobium diselenide, the DNA content actually increased, by as much as 15% (P = 0.8%). The DNA content dropped in the lung tissue under the action of selenium and tungsten and molybdenum diselenides; the administration of niobium diselenide actually brought about an increase of the quantity of DNA, to 152%. The RNA level changed similarly. Three months after the uptake into the organism of the more toxic metallic selenium and metal selenides the content of RNA as a rule dropped in the liver and in some cases in the lungs. The less toxic compounds – niobium, molybdenum, and tungsten diselenides – during the first month of their action increased the RNA level in the lung tissue and liver. The intensified synthesis of RNA and DNA is apparently related with a proliferative reaction in the lung tissue and with regenerative processes in the liver, whereas the decrease of DNA and RNA can be regarded as a consequence of the death of cells and disturbance of the biosynthesis of nucleic acids due to the toxic agents.

The results yielded by studies of the indices of protein and nucleic acid metabolism permit the assumption that the development of the pathological process is related to a considerable extent with affection primarily of the liver and, to a smaller extent, of other parenchymatous organs.

Many authors report disturbance of protein metabolism in various hepatopathies, particularly in toxic hepatitides [70, 79]. According to [79], in these diseases the quantity of albumins decreases and the content of globulins increases. An increase of the content of γ-globulins and a decrease of the albumin–globulin ratio were observed in patients suffering from occupational poisoning by lead, beryllium, and other agents [70, 270]. Such changes occurred at early stages of intoxication, and may be regarded as a nonspecific response of the organism to various stimuli. The content of albumins dropped in the case of pronounced intoxications. The content of transaminases in some patients was elevated.

The morphological disorders induced by metal selenides occur predominantly in the parenchymatous organs. The most pronounced are the alterations in the liver, in which parenchymatous (granular) degeneration develops first and is followed by more severe disorders – fatty degeneration (Fig. 32). Foci of lymphoid cells and histiocytes formed in the parenchyma of the liver in some animals (Fig. 33). There was also a toxic effect on the kidneys. As a rule we noted parenchymatous degeneration of the epithelium of the renal convoluted tubules with necrotic areas, and in certain instances swelling of the capillaries of the renal glomeruli (Fig. 34).

Changes occur also in the structure of the spleen and adrenal glands. Comparatively feeble disorders arise in the lungs of experimental animals, which manifest themselves in the development of bronchitis, peribronchitis, sometimes perivasculitis, and also

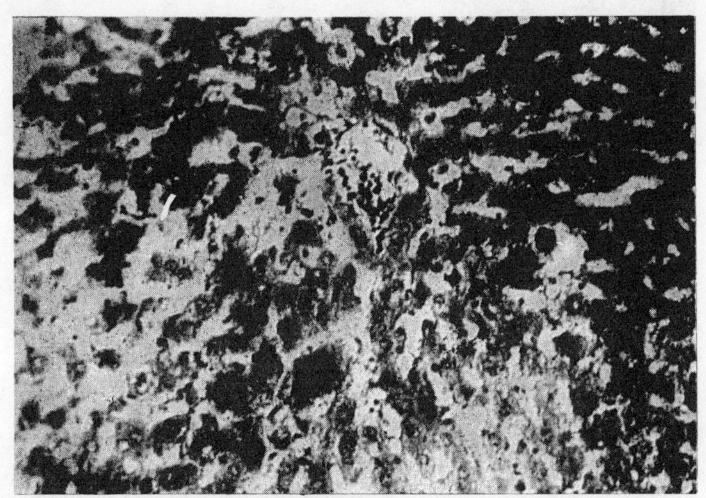

Fig. 32. Severe form of degeneration of the liver six months after endotracheal administration of tungsten diselenide. Magnification 7 × 20.

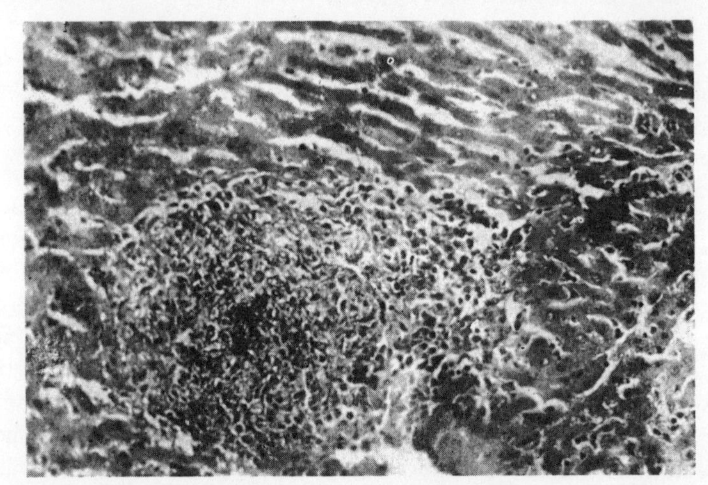

Fig. 33. Histiocytic nodule in rat liver three months after administration of tungsten diselenide. Magnification 7 × 20.

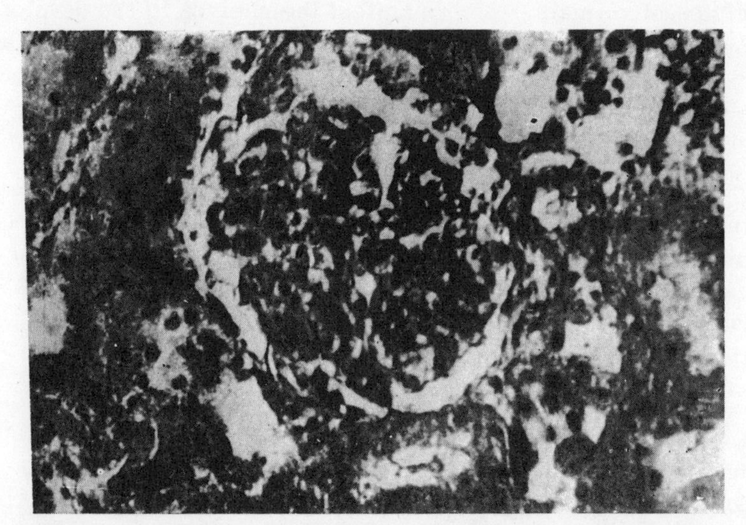

Fig. 34. Degeneration of the epithelium of the renal glomeruli and tubules with necrotic phenomena six months after endotracheal administration of tungsten diselenide. Hematoxylin-eosin staining. Magnification 7 × 40.

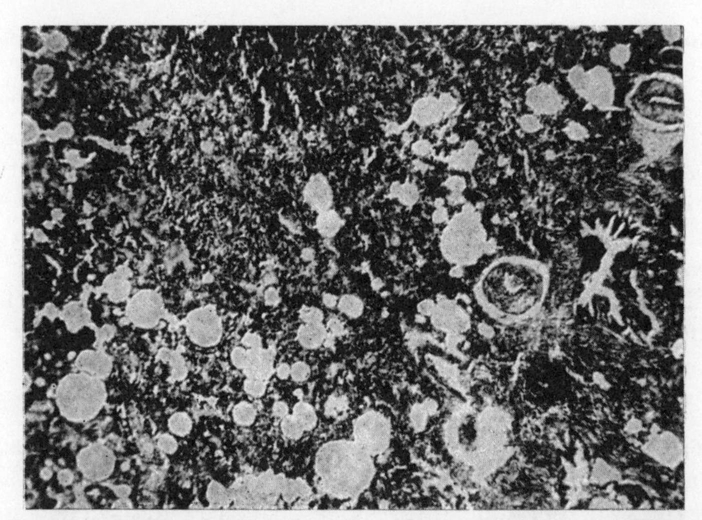

Fig. 35. Bronchopneumonia, deposition of dust particles, and thickening of the pulmonary vessels a month after the administration of tungsten diselenide. Magnification 8 × 10.

focal interstitial pneumonia with the development of cellular-dust nodules at later periods of intoxication (Fig. 35). It must be pointed out that the dust of transition metal selenides is only slowly eliminated from the organism, since even six months after endotracheal administration dust particles were found in the interalveolar septa and peribronchial lymph nodes.

A characteristic feature of the pathological action of selenium and selenides is widespread basophilia of cell elements over appreciable portions of all the organs and tissues investigated, the occurrence of which is apparently related with metabolic disorders, particularly oxidation–reduction reactions induced by the materials studied.

Selenium and its compounds can be arranged in the following order of decreasing morphological alterations: selenium, tungsten diselenide, molybdenum diselenide, and niobium diselenide. Thus, the expressivity of the changes corresponds to the metabolic disorders noted above.

A comparison of data on the effect of selenides, selenium, and transition metals gives reason to assume that the above-described changes in animals that have received selenides are due more to the toxic effect of selenium than that of the metals. Quite appreciable, too, is the effect of the metallic components which to some extent weaken the effect of selenides, in connection with which the toxic effect of selenides is weaker than that of metalllic selenium. The aforesaid is confirmed by results of experiments on inhalation poisoning of animals by tungsten diselenide for six months at a concentration of 1-10 mg/m^3 (average 3.7 ± 0.4). As mentioned earlier, changes occurred in animals exposed to tungsten diselenide at the concentration indicated, providing evidence for the toxicity of these compounds. In view of this, the MAC for tungsten diselenide should be less than the 6 mg/m^3 adopted for tungsten.

Taking into account that the toxic effect of the investigated material is similar to the effect of selenium, for which the MAC is 2 mg/m^3, we recommend the same concentration for tungsten diselenide.

Since molybdenum and niobium diselenides have a less pronounced toxic effect in comparison with the toxicity of selenium and tungsten diselenide, we can recommend 4 mg/m^3 as the MAC for these compounds.

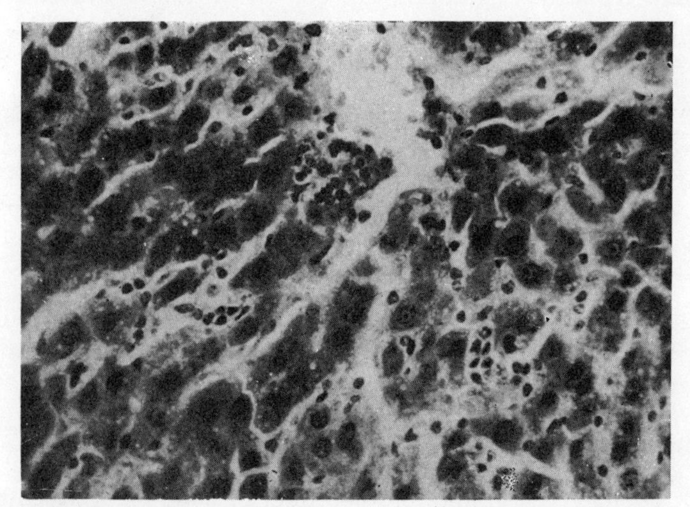

Fig. 36. Edema and degeneration of the liver, proliferation of Kupffer's cells six months after endotracheal administration of molybdenum ditelluride. Magnification 10 × 20.

On studying the effect of the dust of tellurium, molybdenum ditelluride $MoTe_2$, and tungsten ditelluride WTe_2 on the organism, we established that these substances have a considerable toxic effect. When these substances are taken up into the organism collagen formation is intensified and disorders of nucleic acid and protein metabolism occur as a result of blocking of the SH groups of the protein molecules. Intensified oxidation not only of the free but also of the bound SH groups occurs as a consequence of denaturation of proteins. As a result, pronounced pathological changes in the lung tissue and toxic affections of the myocardium, liver, kidneys, and adrenal glands take place. The organs of the reticuloendothelial system react vigorously. This is indicated by disorders in the parenchymatous organs, hyperplasia of the splenic follicles, and marked proliferation of Kupffer's cells of the liver (Fig. 36).

Comparing the changes noted with the pathological process observed under the action of selenium and molybdenum and tungsten diselenides, we can conclude that the tellurides are more toxic than the selenides. It is characteristic that tungsten diselenide is more active than molybdenum diselenide and that tungsten ditelluride is more active than molybdenum ditelluride.

Taking into account that tungsten ditelluride is close in toxicity to tellurium, we recommend the same MAC for tungsten ditelluride as for tellurium, i.e., 0.01 mg/m^3. For molybdenum ditelluride, it can be slightly higher, i.e., 0.1 mg/m^3.

Generalizing the data on the toxicity of compounds belonging to the chalcogenide class, we can conclude that their biological activity increases from the sulfides to the selenides and then to the tellurides.

CHAPTER VII

Toxic Effect of Transition Metal Carbonyls

Nickel carbonyl was the first of the carbonyls to be discovered (by Mond in 1890). At present the carbonyls of many metals and of some nonmetals are produced. The carbonyls of the transition metals of Groups VI–VIII of the periodic system have been studied most.

In chemical structure metal carbonyls are complex compounds for which the number of ligands (coordination number) exceeds the usual valence of the central atom. The chemical bonds in the majority of molecules are covalent. Their number is determined by the usual valence of the atom, which is constant in the sense that the number of unpaired valence electrons is governed by the structure of the electron shell.

The structure of metal carbonyls is interpreted in the light of Sidgwick's theory of the structure of complex compounds, according to which each CO molecule added to the metal gives it two additional outer electrons. As many CO molecules are added to the metal atom as necessary for the effective atomic number of the central atom to become equal to the atomic number of the inert gas following it. To acquire the electronic structure of the inert gas argon, the iron atom must transfer eight 3d4s electrons, and to form an electron shell analogous to the inert gas krypton, the iron atom must externally acquire 10 electrons.

Five carbon monoxide molecules are added directly to the central iron atom by means of two 2s electrons of the carbon atom. Nickel in carbonyl compounds adds four carbon monoxide molecules, which form four coordination bonds due to the 2s electrons of the carbon atoms of the gas and four 4s4p cells of the metal atom [24, 246].

It was established on the basis of numerous investigations that in a nickel carbonyl molecule carbon monoxide has the same structure as in a free state; the double bond N : : C is present.

Molybdenum, having the electronic configuration d^5s^1, in order to acquire the structure of the inert gas xenon s^2p^6 must accept 12 electrons or, building up to the configuration of krypton $4s^2p^6$, lose six electrons.

In compounds with a covalent coordination bond the molybdenum atom can acquire only the electronic structure of xenon, since it participates with six free $4d^55s^1$ electrons and three free 5p cells in the formation of the carbonyl; therefore molybdenum carbonyl has the formula $Mo(CO)_6$.

The tungsten atom d^4s^2 needs 12 electrons to acquire the electronic configuration of the inert gas radon. Tungsten carbonyl with a covalent bond has the formula $W(CO)_6$.

The transition metal carbonyls are formed upon interaction of d electrons, the atoms of carbon and the metal being interlinked through the electron pair represented by the CO group.

The carbonyls of these metals have various spatial structures: $Ni(CO)_4$ a tetrahedral structure, $Cr(CO)_6$ octahedral, and $Fe(CO)_5$ a structure of a trigonal bipyramid. Tungsten carbonyl $W(CO)_6$ forms an orthorhombic structure. The CO molecules surround the metal atom in the form of a regular octahedron, the carbon atom being arranged along a straight line between the metal and oxygen atoms.

The majority of metal carbonyls are crystalline substances, and only the carbonyls of nickel, iron $Fe(CO)_5$, ruthenium $Ru(CO)_5$, and osmium $Os(CO)_5$ are liquids. Metal carbonyls are diamagnetic, sometimes having only a small dipole moment, readily soluble in organic solvents, and very volatile [18, 246].

Chemically, metal carbonyls are distinguished by high reactivity. The carbonyls of alkali metals ignite in air, are insoluble in organic solvents, and have pronounced polarity. Carbonyls of the transition metals are more stable. On heating above a certain temperature metal carbonyls decompose with the formation of carbon monoxide and a finely dispersed metal.

The glistening white crystals of tungsten hexacarbonyl volatilize at 50°C and decompose (beginning at 100-150°C) with the formation of a metallic surface and liberation of blue tungstic oxide. The crystals sinter at 125°C.

The presence of fragments (intermediate decomposition products) of $W(CO)_4$, $W(CO)_3$ (numerous), $W(CO)_2$, and $W(CO)$ is detected upon thermal dissociation of tungsten carbonyl in a mass spectrometer.

Despite its chemical inertness, tungsten carbonyl enters into various substitution reactions with certain amines, forming derivatives with a smaller number of CO groups, for example, with pyridine – $W(CO)_3C_6H_5N$; $W(CO)_4(C_6H_5N)_2$.

Metal carbonyls are produced by the action of CO on the metals at high pressures and temperatures, and also by the reaction of CO with metal chlorides, halides, or sulfides.

From a technical standpoint, nickel, cobalt, and iron carbonyls are the most important. The carbonyls are used for producing pure metals by thermal decomposition. Iron, cobalt, and nickel carbonyls are used as catalysts of important chemical processes; they are good antiknock agents of gasoline, but difficultly removable oxides are formed upon their burning. Some carbonyls are used for producing high-purity carbon monoxide.

Toxic Properties of Carbonyls. Various disorders can occur in workers in the production and industrial use of carbonyls. Cases of acute and chronic carbonyl poisoning are described in the literature. As was indicated in Chapter III, the entry of nickel carbonyl via the skin or inhalation of its vapors caused in workers symptoms of affection of the nervous system, pulmonary edema, cardiovascular disorders, hepatic lesions, irritation of the air passages, and other phenomena. Fatal cases of poisoning have been reported in which the clinical state resembled symptons of poisoning by suffocating gases.

An experiment on animals with nickel carbonyl poisoning established a decrease of oxygen consumption even before the development of pulmonary edema [167]. Inhalation of nickel carbonyl vapors for 30 min killed 50% of mice at a concentration of 0.07-1 mg/liter, and of white rats at 0.2-0.4 mg/liter [167]. The lethal concentration for cats and dogs with 30-min inhalation of nickel carbonyl vapors is 2-2.5 mg/m^3. A histological investigation of the internal organs of the experimental animals revealed pulmonary edema, multiple hemorrhages, including some in the brain tissues, parenchymatous and fatty degeneration of the epithelium of the renal convoluted tubules, and degenerative changes of the incretory and excretory parts of the pancreas and the adrenal glands.

Inflammatory changes, purulent bronchitis, and metaplasia of the bronchial epithelium developed in chronic poisoning by nickel carbonyl vapor. The animals gradually lost weight, and lung cancer was found in two of 10 animals. The nickel content in the lungs and especially in the kidneys of these animals was higher than in the control [167, 568, 573, 574].

Iron carbonyl $Fe(CO)_5$, which is a liquid, enters the organism via undamaged skin and air passages and causes pulmonary edema. The toxicity of iron carbonyl is almost half that of nickel carbonyl; its LD_{50} for mice is 2.19 mg/liter and for white rats 0.91 mg/liter [368].

The mechanism of the toxic action of nickel carbonyl consists in the following: Entering the organism, it dissociates with formation of carbon monoxide and nickel ions. The latter, forming complexes with proteins, enter the blood, then the brain, liver, kidneys, and heart, and bring about pathological disorders [314].

Less toxic is cobalt carbonyl $CO_2(CO)_8$, which, applied to the skin, causes depilation and dry scabs which subsequently heal despite continued contact with the compound [325].

The liquid cobalt carbonyl hydride $Co(CO)_4H$, which in a concentration of 0.165 mg/liter killed 50% of experimental white rats within 30 min, has a greater toxicity. Some of the animals died after chronic inhalation for three months. An increase of the hemoglobin level and a decrease of the number of lymphocytes and monocytes were observed after six months. Cellular-dust nodules with fibrotic phenomena and inflammation, which disappeared after six months, were noted in the lungs of the dead rats.

Up to 0.31 mg of Co was found in the lung tissues. About 15-12% of the quantity of carbonyl hydride inhaled was excreted in the urine per day [161].

There are no data in the literature on the effect of molybdenum carbonyl on the organisn. The toxic effect of tungsten carbonyl is indicated in Frolova's report [362]. Experiments on white rats and rabbits established that tungsten carbonyl does not have a toxic effect when applied to the skin and administered into the stomach in doses of 1.5-5 mg/kg. Repeated oral administration of $W(CO)_6$ in a dose of 0.5 g/kg for three months caused a lag in weight gain. Tungsten carbonyl inhaled once in a concentration of 0.35-0.45 mg/liter produced no toxic effect; the repeated inhalation of tungsten carbonyl in these same concentrations for 40 days caused convulsions.

Delamination and increased fragility of the nails and desquamation of the skin were noted in workers exposed to tungsten carbonyl for a long time.

In view of the lack of data on the effect of molybdenum carbonyl on the organism and limited information on the toxicity of tungsten carbonyl, we carried out special investigations to determine the fibrogenic and general toxic effects of these compounds.

The experiments were conducted on 90 white rats weighing 120-150 g each. The substances investigated – fine divided powders – were administered to the animals endotracheally for one, three, and six months.

The condition of the animals during the entire experimental period did not differ from the control animals. They all gained weight.

The collagen content increased slightly after the administration of molybdenum and tungsten carbonyl, whereas the quantity of ascorbic acid did not change appreciably. The collagen content three months after poisoning with molybdenum carbonyl increased by 37.7% and by 20.9% after the administration of tungsten carbonyl (Table 26).

The content of RNA in the lung tissue one and six months after the administration of molybdenym carbonyl decreased 39.7-34.4% ($P < 0.05$) in comparison with the control, whereas after

TABLE 26. Change of Collagen Content in Lung Tissue of White Rats after Endotracheal Administration of Molybdenum and Tungsten Carbonyls

Substance administered	Exptl. period, months	Content, mg	Increment of collagen	
			mg	%
Molybdenum carbonyl	1	25.9±1.33	1.5	10.4
	3	32.6±3.2	8.2*	37.7
	6	22.6±2.3	—3.7	—
Tungsten carbonyl	3	29.5±1.8	5.1*	20.9
	6	31.9±3.1	5.6	21.2

*$P < 0.05$; in other cases $P > 0.05$.

three months we observed a statistically significant increase of the RNA content by 17% (Table 27). The DNA content in the lung tissue of such animals during the entire experiment was lower than in the control, but a statistically significant decrease by 31.5% was noted only after three months.

The level of RNA and DNA in the lung tissue of the animals six months after the administration of tungsten carbonyl was higher but this increase was not statistically significant ($P > 0.05$).

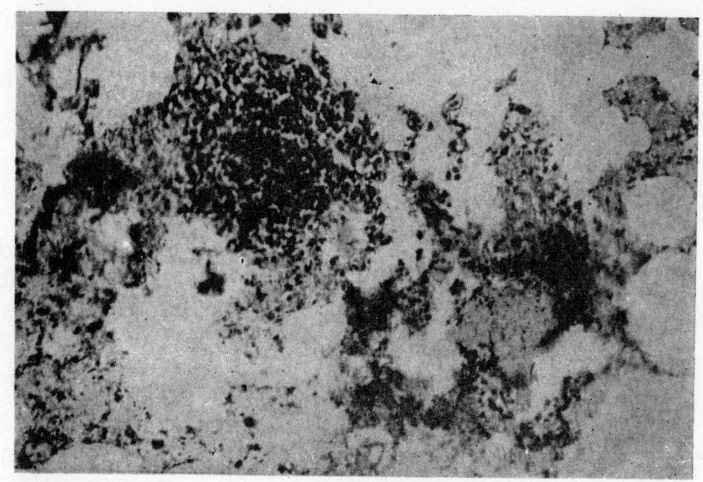

Fig. 37. Cellular-dust nodule in lung tissue a month after endotracheal administration of molybdenum carbonyl. Magnification 7 × 10.

TABLE 27. Change of Nucleic Acid Content in Lung Tissues and Liver of White Rats after Endotracheal Administration of Molybdenum and Tungsten Carbonyls

Substance administered	Exptl. period, months	In lungs						In liver					
		RNA			DNA			RNA			DNA		
		Content, μg P	Increment		Content, μg P	Increment		Content, μg P	Increment		Content, μg P	Increment	
			μg P	%		μg P	%		μg P	%		μg P	%
Molybdenum carbonyl	1	153.9±4.6	–98.7*	–39.7	344±67.3	–114	–24.9	231.2±98.6	–124	–34.8	337.8±104	–86.6	20.4
	3	295±37	43*	17.0	314±1.17	–149*	–31.5	343±64	12	3.3	307±8.5	–118,7*	–28
	6	175±6,8	92*	–34.4	445.6±20,4	–0.5	–0.19	280±56	4	1.4	437,8±87.5	104,8*	31.5
Tungsten carbonyl	6	280±17.3	12	44.9	469.5±1.5	23,4	5,2	361±75,4	76.8	26,9	474.5±142	14.2	42,9

*$P < 0.05$; in other cases $P > 0.05$.

The morphological changes in the lung tissue of the animals a month after the administration of molybdenum carbonyl was characterized by pronounced hyperplasia of the lymphoid tissue around the bronchi and vessels and by thickening of the inter-alveolar septa. Small, round cellular-dust foci were noted in some animals at sites of dust-particle aggregates (Fig. 37). In addition to these disorders, development of a fibrotic process could be noted in the bronchial walls at later periods.

Moderate parenchymatous degeneration of the myocardium with phenomena of toxic edema and proliferation of the endothelium of the capillaries developed within one month after the administration of molybdenum carbonyl. Phenomena of cloudy swelling were detected in the liver. In many areas of the preparations granulation and cloudy swelling were noted in the protoplasm of the liver cells; the outlines of the cells were weakly differentiated, and the nuclei of some cells were faintly stained or pycnotic, sometimes in a state of mitosis; binuclear cells were found, which indicated phenomena of regeneration of the hepatic parenchyma. At the same time focal parenchymatous degeneration of the epithelium of the renal convoluted tubules and hyperplasia of the splenic follicles developed.

After three and six months the degenerative changes in the internal organs were even more pronounced. The muscle fibers of the myocardium were stained unevenly and swollen; cloudiness of the protoplasm, fragmentation, and occasional marked edema were observed. Signs of fatty degeneration were noted in the liver together with the phenomena of cloudy swelling. On preparations stained with hematoxylin eosin we noted numerous vacuoles located intra- and extracellularly. Fat inclusions were observed at these sites on staining with Sudan III. The changes in the renal parenchyma were more marked.

Perl's test for iron demonstrated more blue inclusions in the splenic red pulp than in the control animals, which indicated increased disintegration of erythrocytes.

Degenerative changes, but less pronounced, also occurred in the parenchymatous organs of the experimental animals that had received tungsten carbonyl. Degenerative changes were found in the myocardium after a month and especially after three months. The muscle fibers were stained unevenly and basophilia of the protoplasm was noted; sometimes fragmentation and swelling of the muscle fibers were detected.

Phenomena of cloudy swelling and in some instances fatty degeneration were observed on investigating histological sections of the liver. Swelling of the epithelium of the convoluted tubules, sometimes with phenomena of parenchymatous degeneration occurred in the kidneys. Restricted aggregations of lymphocytes were found sometimes in the renal interstices.

Moderate hyperplasia of the red pulp was note in the spleen after three and six months. Perl's test for iron showed the presence of a considerable number of blue inclusions.

After six months we noted in the kidneys diffuse parenchymatous degeneration, edema of the glomeruli, perivascular edema, and an albuminous exudate in the cavity of the capsules.

The toxic effect of the substances investigated is indicated also by the change of certain biochemical indices of the state of the liver. The RNA content in the liver tissue during the six months did not change appreciably in the experimental animals (Table 27). The quantity of DNA in the liver tissue three months after the administration of molybdenum carbonyl decreased by 28%

TABLE 28. Change of Content of Aminotransferases in Blood Serum of White Rats after Endotracheal Administration of Molybdenum and Tungsten Carbonyls

Substance administered	Exptl. period, months	Aspartate aminotransferase			Alanine aminotransferase		
		Content, arbitrary units	Increment		Content, arbitrary units	Increment	
			Arbitrary units	%		Arbitrary units	%
Tungsten carbonyl	3	143.7±7.2	106.8*	291.8	39.9±1.4	25.5*	177.5
	6	39.7±1.05	5.9	12.9	31.1±7.2	0.5	1.6
Molybdenum carbonyl	1	21.4±2.2	—7.2*	24.2	15.2±3.9	−13.4*	46.8
	3	151.7±3.4	114.8*	311.1	38.6±2	24.2*	166.5
	6	49.6±6.1	4	8.7	36.5±11.9	5.9	19.2
Physiological solution (control)	1	29.7±1.3	—	—	28.6±1.14	—	—
	3	36.9±0.67	—	—	14.4±1.0	—	—
	6	45.57±3.2	—	—	30.6±2.25	—	—

*$P < 0.05$; in other cases $P > 0.05$.

in comparison with the control ($P < 0.05$) and after six months increased by 31.5% ($P < 0.05$).

An increase of the DNA content in the liver was noted also six months after administration of tungsten carbonyl. However, the data were statistically insignificant.

The phasic changes of the content of nucleic acids induced by molybdenum carbonyl are apparently related with disturbance of the morphological structures of the liver cells. Three months after administration of the material we observed maximum destructive changes in the hepatic parenchyma. The increase of the DNA content after six months was probably due to intensification of regeneration of the tissue [125, 255]. Data from a study of the activity of aminotransaminases in the blood serum also indicate toxic affection of the liver.

A month after the administration of molybdenum carbonyl we found a slight decrease of activity of both enzymes, and after three months a sharp increase of the activity of aspartate and alanine aminotransferases ($P < 0.05$). The increment of aspartate aminotransferase was 311.1% in comparison with its level in the control animals, and of alanine aminotransferase, 166.5%. By six months the activity of the enzymes had also returned to the normal level (Table 28).

The maximum increase of activity of the enzymes after adminstration of tungsten carbonyl was also noted after three months. For aspartate aminotransferase it was 291.8% in comparison with the control and for alanine aminotransferase 177.5%. By the sixth month the activity of both enzymes had returned to the level of the control animals. The results of this series of experiments also indicate a toxic affection of the liver, although the possibility of a change of activity of aspartate aminotransferase in connection with degenerative phenomena in the myocardium cannot be ruled out.

Our investigations showed that molybdenum and tungsten carbonyls do not cause acute poisoning, but can be the cause of chronic intoxication.

The pathological states occurring are due to the effect of the metallic components of the compounds. The pathological changes in the internal organs are similar to those induced by pure metals and some refractory compounds. Intensification of the toxic effect is apparently due the effect of carbon monoxide. Localization of the pathological alterations corresponds to the reticuloendothelial type of distribution characteristic of metals [316, 317]. Therefore, we can assume that molybdenum and tungsten carbonyls, on entering the organism, break down into CO groups and metal ions. The latter probably form complexes with proteins and nucleic acids, thereby causing disorders of protein metabolism and enzymatic processes in the organism.

The biological activities of molybdenum and tungsten carbonyls are not the same. Molybdenum carbonyl has greater toxicity. Tungsten carbonyl causes milder inflammatory phenomena in the lung tissue and predominantly cloudy swelling in the liver and kidneys, whereas fatty degeneration (a more severe condition) develops in the liver after the administration of molybdenum carbonyl.

The lower toxicity of tungsten carbonyl is indicated by feeble disturbances in nucleic acid metabolism and smaller intensity of the biosynthesis of collagen. A comparison of the toxic effect of these compounds with the literature data on the toxicity of other carbonyls indicates that the biological activity of carbonyls decreases from nickel carbonyl to iron carbonyl, cobalt carbonyl hydride and cobalt carbonyl, then to molybdenum and tungsten carbonyls.

The observed regularity in the change of the toxicity of carbonyls can apparently be explained by the characteristics of their electronic structure.

As the maximum allowable concentration we can recommend 1 mg/m^3 for molybdenum carbonyl and 2 mg/m^3 for tungsten carbonyl.

CHAPTER VIII

Theoretical Problems of the Toxicology of Metals and Their Compounds

In connection with the development of chemistry and the ever increasing demands of technology, various metals and metallic compounds and alloys are being introduced into industry which under production conditions can have a toxic effect on the workers. Therefore, there is an urgent need to elucidate the general regularities of the action of these substances on the organism in order to predict their toxicity and to determine the maximum allowable concentrations in the air of industrial rooms.

Development of concepts concerning the nature of the chemical bond enables us to use, for analyzing the causes and determining the level of toxicity of substances, data on the electronic structure and stability of chemical bonds occurring between atoms upon formation of inorganic substances [2, 16, 256, 307, 337].

Some authors [261, 273, 318, 367] examined the degree of biochemical activity of a number of substances as a function of their electronic structure and made appropriate quantum mechanical calculations. Approaches to the determination of toxicity and MAC by calculation based on an analysis of the correlation between biological activity and various physicochemical properties of many organic substances have been developed by Soviet and foreign authors [158, 179, 181, 358, 506].

A high degree of correlation of such properties in organic substances with their MAC has also been established. The MAC is an integral index reflecting many aspects of the deleterious effect of substances and is important in a practical respect.

Establishment of the relation between toxicity and electronic structure characteristics is the most promising direction of investigations aimed at calculating the toxic properties and MAC for metals and their compounds.

Few studies have been devoted to the establishment of a relation between the toxicity of a metal ion and the position of the element in the periodic system.

In 1955, Saccardo [546] suggested that all the elements of the periodic table could be arranged according to their toxicity along a spiral (from the periphery toward the center), their position being given by the intersection of the spiral with 16 radii. However, since he had to make many exceptions, the author has concluded that there is no correlation between the degree of toxicity and the internal structure of the atom, as reflected in the periodic system.

Bienvenu, Nofre, and Cier [394] established LD_{50}, i.e., the doses killing 50% of animals in 30 days, for the cations of soluble compounds of 41 elements and came to the conclusion that the periodicity of the changes of toxicity of metal ions is related with the electronic structure of the atom of the element, like certain physicochemical characteristics of simple substances. The toxicity of an element increases with an increase of atomic number, but in the transition metal groups the opposite relation is observed.

It was established in [517] that toxicity is variously related with the valence of an anion. It increases with increase of valence for halogen and phosphorus compounds, but decreases for chalcogens, nitrogen, and arsenic; this regularity is apparently due to different donor–acceptor properties of the anions.

Changes of toxic properties are found both in nonmetals and in metals, which can probably explain the difference in the toxicity of the same element with an increase of valence upon the formation of compounds with different anions. However, the authors failed to allow for this and consequently found it difficult to establish definite, regular relations between the biological activity of compounds and the position of the atoms of their components in the periodic systen.

The existence of relations between the toxicity of metal ions and various constants of elements and their compounds (sulfides and oxides) was confirmed in [181] on the basis of extensive data. A relation between toxicity and stability of oxygen compounds of higher valence, and also ionization potentials and atomic radii, was established. Since the latter are a periodic function of the atomic number of the elements, the author concludes that a relation exists between the toxic properties and atomic structure of a given element.

For confirmation of this a comparison was made between the toxicity and electronic structure of two groups of simple substances having minimum toxicity – inert gases and elements of the first group (Li, Na, K, Rb, Cs) – and maximum toxicity – Cu, As, Se, Cd, In, Fe, Pt, Hg, U. It was concluded on the basis of the results obtained that the least toxic are those elements whose electron shells are completely filled by electrons (inert gases, and also elements of the first principal subgroup with one electron in the outermost shell). The most toxic, in the author's opinion, are elements with unfilled electron shells.

In [183] the author confirmed the aforementioned correlation between the toxic effect of elements and their position in the periodic system. Using the data of Bienvenu et al. [394] and of other investigators on LD_{50}, the author compared the logarithms of these doses with the position of the elements in the periodic system and confirmed the periodic character of change of toxicity. The recurrence of the change of toxicity by periods was established in a graphic representation. In each large period there are three peaks, of which the first and highest corresponds to the toxicity of inert gases and the third and lowest corresponds to the most toxic elements. The toxicity of the elements of the middle period is characterized by a straight line drawn through the pertinent points of the first and third periods. The LD_{50} values of the elements of the second large period occupied the middle position between the toxicity of the elements of the first and third large periods. The author points to a peculiarity of the change of toxicity in the secondary subgroups of Groups I, V, and VI, in which toxicity decreases on passing from a metal with a smaller atomic number to an element with a higher number ($V < Nb < Ta$; $Cr < Mo < W$; $Cu < Ag < Au$).

These data indicate a definite relation between the toxic effect and position of elements in the periodic system, and also a correlation between toxic and physicochemical properties. An attempt was also made to compare the data on the toxicity of the most and least active elements with their electronic structure.

To work out suitable approaches to a determination of the degree of toxicity of substances, simultaneously with the aforementioned investigations we compared the toxic properties of various simple substances, metallic oxides, a number of refractory compounds, chalcogenides, and metallic carbonyls with their electronic structure [36, 38, 39, 46, 49].

Unlike the earlier studies, our investigations were based on concepts of the electronic structure of the substances studied in relation to the probability of the formation of stable configurations by their atoms within the framework of the configurational localization model, and also with allowance for the type of chemical bond formed and crystallochemical properties.

Variation of the Toxic Properties of Simple Substances as Function of Their Electronic Structure

Concepts of stable electronic configurations and their role in the formation of the properties of chemical elements and compounds constitute the general principle on the basis of which the nature of the toxic effect of simple substances is examined in relation to their electronic structure [303, 307, 308].

To evaluate the character of the general effect of substances on the organism as a function of their electronic structure, we used results of our own observations and literature data on symptoms of intoxication, LD_{50} [394], and MAC [313] in current use (Table 29). For convenient presentation of the material it is best to examine the data indicated by dividing the elements into three groups according to their electronic structure:

1. Alkali and alkaline earth metals, and also metals of the copper and zinc groups – s metals;
2. Elements whose inner electron d and f shells are being completed, so-called sd and fsd elements, which include all transition metals;
3. Elements with outer sp electrons, i.e., nonmetals and semimetals.

TABLE 29. Electronic Structure of Elements and Maximum Allowable Concentrations of Various Substances, mg/m^3

Period	I	II	III	IV	V	VI	VII	VIII		
	A B	A B	A B	A B	A B	A B	A B	A	B	
1	1 H s^1							2 He s^2		
2	3 Li s^1	4 Be s^2 0,001	5 B s^2p^1	6 C s^2p^2 10,0	7 N s^2p^3 N_2O_5—5,0	8 O s^2p^4	9 F s^2p^5	10 Ne s^2p^6		
3	11 Na s^1 0,5 (NaOH)	12 Mg s^2	13 Al s^2p^1 2,0	14 Si s^2p^2 SiO_2—1,0	15 P s^2p^3 0,03	16 S s^2p^4	17 Cl s^2p^5	18 Ar s^2p^6		
4	19 K s^1	20 Ca s^2	21 Sc d^1s^2	22 Ti d^2s^2 TiO, TiO_2,—10,0	23 V d^3s^2 0,1—0,5	24 Cr d^3s^1 Cr — 2,0 CrO_3 — 0,01	25 Mn d^5s^2 MnO_2 — 0,3	26 Fe d^6s^2	27 Co d^7s^3 0,5	28 Ni d^8s^2 0,5
	29 Cu $d^{10}s^1$ 0,5	30 Zn $d^{10}s^2$ 5,0	31 Ca s^2p^1	32 Ge s^2p^2 2,0	33 As s^2p^3 As_2O_3 — 0,3	34 Se s^2p^4 2,0	35 Br s^2p^5	36 Kr s^2p^6		
5	37 Rb s^1	38 Sr s^2	39 Y d^1s^2	40 Zr d^2s^2 5,0	41 Nb d^4s^1 10,0	42 Mo d^5s^1 2—4,0	43 Tc d^5s^2	44 Ru d^7s^1	45 Rh d^8s^1	46 Pd $d^{10}s^0$
	47 Ag $d^{10}s^1$	48 Cd $d^{10}s^2$ CdO — 0,1	49 In s^2p^1	50 Sn s^2p^2 5,0—6,0	51 Sb s^2p^3	52 Te s^2p^4 0,01	53 I s^2p^5	54 Xe s^2p^6		
6	55 Cs s^1	56 Ba s^2	57 La d^1s^2	58—71 Lan-thanides 72 Hf d^2s^2	73 Ta d^3s^2	74 W d^4s^2 6,0	75 Re d^5s^2	76 Os d^6s^2	77 Ir d^7s^2	78 Pt d^9s^1
	79 Au $d^{10}s^1$	80 Hg $d^{10}s^2$ 0,01	81 Tl s^2p^1	82 Pb s^2p^2 0,01	83 Bi s^2p^3	84 Po s^2p^4	85 At s^2p^5	86 Rn s^2p^6		
7	87 Fr s^1	88 Ra s^2	89 Ac d^1s^2	90 Th d^2s^2 0,05	91 Pa	92 U d^1s^2 0,015 0,075				

We will first consider the sp elements (nonmetals and semimetals) for which the most complete and systematic data have been obtained.

It is known that carbon, found in nature as amorphous coal, graphite, and diamond, and elemental silicon are relatively inert. The maximum allowable concentration for coal dust is 10 mg/m^3. The MAC has not been established for silicon, but, judging by the character of its pathological effect, it should be close to that indicated. Both types of dust effect predominantly a local reaction, characterized by aggregation of alveolar macrophages and feeble pulmonary fibrosis.

The effect of germanium dust on experimental animals, in addition to changes of an inflammatory-proliferative character in the lung tissue, causes retardation in weight, degenerative changes in the parenchymatous organs, and other phenomena attesting to the toxicity of the substance. The MAC for germanium dioxide is 2 mg/m^3 [313]. The administration of tin leads to pneumoconiotic changes in the lung tissue, which at later periods are characterized by pronounced changes of the bronchial walls with the formation of collagen fibers at the affected sites. The MAC for tin is 5-6 mg/m^3 [366]; LD_{50} for Sn^{2+} is 0.346 mg-atom/kg.

Lead, as is known, has a pronounced toxic effect. Its MAC is 0.01 mg/m^3 and LD_{50} for Pb^{2+} is 0.370 mg-atom/kg. On entering the organism it binds the SH groups, thereby inducing pathological changes, mainly in the blood, nervous and cardiovascular systems, and parenchymatous organs.

A comparison of the biological activity of these elements indicates an increase of toxic properties from carbon to lead. The atoms of these elements in an isolated state have the electronic configuration s^2p^2 which changes to the energetically more stable configuration sp^3* upon formation of a solid, owing to $s \rightarrow p$ transition.

In diamond only sp^3 configurations are formed, which provide a tetrahedal coordination of the atoms, and in graphite sp^2 configurations (the change of diamond to graphite is related with the destruction of the sp^3 configuration: $sp^3 \rightleftharpoons sp^2 + p$). Since the energy stability of similar electronic configurations decreases

*The sp configurations can be arranged according to increasing energy stability in the series $s^2p \rightarrow sp \rightarrow sp^2 \rightarrow sp^3$.

with an increase of the principal quantum number of valence sp electrons [308], for carbon the $s^x p^y$ configurations are the most stable, and considerable energy expenditures are required to disturb them. Obviously, therefore, the toxicity of all allotropic modifications of carbon is a minimum. On passing to other elements of this group the electronic configurations of the atoms become increasingly less stable and a part of the valence electrons changes to a delocalized state. The portion of delocalized electrons, which are involved comparatively easily in any processes, governs the chemical activity of the substance. Therefore, whereas silicon and germanium still retain diamond-like structure (based on sp^3 configurations), in tin, in addition to the diamond-like structure (gray tin), a different structural modification (white tin) occurs in which the statistical weight of the sp^3 configurations is negligible (and insufficient for formation of the diamond-like structure).

The process of the decrease in the number of stable electronic configurations comes to an end with lead, which, as far as its external properties are concerned, is a metal (degenerate semiconductor). This apparently also gives rise to its high toxicity.

Two types of changes of electronic configurations are possible for the elements of the nitrogen group (nitrogen, phosphorus, arsenic, antimony, bismuth), which have the electronic configuration s^2p^3 in an isolated state. The first consists in attracting electrons of the partner with the formation of the most energetically stable configuration s^2p^6 (characteristic of inert gases beginning with neon); the second involves giving up one electron with the formation of the configuration of the carbon atom sp^3 according to the scheme $s^2p^3 \rightarrow sp^4 \rightarrow sp^2 + p$. In the first case nitrogen and its analogs should display acceptor properties and in the second, donor. The realization of one or the other possibility depends on the character of the electronic structure of the atom of the partner, its donor–acceptor properties. However, in both cases the toxicity of nitrogen would be expected to be considerably higher than that of carbon as a consequence of its easy ability to accept or transfer electrons, and this is in fact observed (for example, for nitrogen oxides). Molecular nitrogen should be chemicallly inert and nontoxic, since it consists of two stable sp^3 groups interlinked by a strong paired electron bond ($sp^3 : sp^3$).

Indeed, if we turn to data on toxicity, we find that nitrogen (N_2) under ordinary conditions is a physiologically indifferent gas.

On passing from nitrogen to phosphorus the toxicity increases sharply, especially in comparison with the toxicity of molecular nitrogen. White (or yellow) elemental phosphorus is strongly toxic; red is not very toxic because of its weak solubility in body fluids; however, upon entry in the form of dust it can have a toxic effect. The MAC for white phosphorus is 0.03 mg/m^3.

Toxicity increases sharply on passing to arsenic owing to a marked weakening of the stability of the sp^3 configurations and the possibility of simultaneous transfer of one electron as a result of the electronic regrouping indicated above.

For antimony we need take into account the occurrence in its atoms of a completely vacant 4f level, to which can be attracted a part of the sp electrons, in particular, the mobile electron given up on the regrouping of the s^2p^3 configuration to the sp^3. Apparently, therefore, the toxicity of antimony is considerably less than that of arsenic. It can cause acute poisoning, related primarily with the irritating effect of its compounds on the mucous membranes of the air passages, digestive tract, and skin. A general toxic effect, which is especially noticeable in chronic poisoning, is manifested in metabolic disorders, effection of the nervous and cardiovascular systems, change of the morphological composition of the blood, and in degenerative changes of the parenchymatous organs.

For bismuth the number of electrons capable of passing to the 5f level, which is completely free in an isolated atom, is greater than for antimony. This is related with an increase of energy stability of the f state upon an increase of the principal quantum number of the d and f electrons. Cases of occupational poisoning or skin diseases when working with bismuth are unknown so far.

Elements of Group VI, oxygen and chalcogens (sulfur, selenium, tellurium), whose atoms in an isolated state have the electronic configuration s^2p^4, are largely able to capture electrons of their partner with the formation of the stable s^2p^6 configuration. This gives rise to the high chemical activity of the elements of this group, which as a rule is realized by accepting electrons of partners. According to these concepts concerning the stabilization of electronic configurations, the elements in question should have considerable toxicity with a selective effect for those processes in which toxicity is due to electron capture. On the whole, however, the toxicity of these elements should be slightly less than that of the elements of the nitrogen group, since in the latter it can be related with the donation of electrons, and in the chalcogens only with their acceptance.

Oxygen, sulfur, selenium, and tellurium have a pronounced irritating and general toxic effect. Pulmonary edema, dysfunction of the nervous and cardiovascular systems, and others are possible.

For the nonmetals and semimetals of Group III (boron, aluminum, gallium, indium, and thallium), in accordance with their electronic structures (s^2p or sp^2 owing to $s \rightarrow p$ transition) we should expect a high capacity for capuring electrons and a greater toxicity than for elements of the carbon group. Under these conditions the toxicity should increase with increase of the principal quantum number of the sp electrons, i.e., from boron to thallium, and this is in fact observed.

Elemental boron is comparatively inactive, although it sometimes causes sclerosis and nodular changes in the lung tissue and other phenomena. The LD_{50} for aluminum (Al^{3+}) is 0.800 mg-atom/kg, for gallium (Ga^{3+}) 0.210, for indium (In^{3+}) 0.224, and for thallium (Tl^{+}) 0.100 mg-atom/kg.

Toxicity would be expected to increase even more on passing to beryllium, which is actually observed. The LD_{50} for Be^{2+} is 0.150 mg-atom/kg and its MAC is 0.001 mg/m^3. Beryllium has a toxic effect, bringing about marked nodular infiltration of the lung tissue and in individual cases chronic pulmonary granulomatosis. The effect of beryllium dust or fumes on the mucosa of the eyes is often in the form of catarrhal conjunctivitis or chemical burn. The high toxicity of beryllium is related with the fact that its atom, having the electronic configuration s^2 in an isolated state, as a result of $s \rightarrow p$ transition (which is quite probable in view of the low principal quantum number of the valence electrons of beryllium) can acquire the sp configuration – one of the least stable energetically and capable both of building up to more stable states and of disintegrating. The dual nature of beryllium is clearly reflected in its chemical properties.

The probability of sp transitions for magnesium decreases sharply in comparison with beryllium. Its toxicity drops at the same time (LD_{50} of Mg^{2+} is 3.59 mg-atom/kg). In the alkaline earth metals (calcium, strontium, barium, radium) the s^2 configurations can be disturbed with transfer of part of the electrons to the completely vacant d orbitals ($s \rightarrow d$ transitions). Here the probability of $s \rightarrow p$ transitions decreases considerably, but the toxicity increases slightly in comparison with that of magnesium owing to the appearance of s^1 electrons, which are able to capture electrons of partners with the formation of s^2 configurations. The LD_{50} for Ca^{2+} is equal to 2.50, for Sr^{2+} 5.73, and for Ba^{2+} 0.258 mg-atom/kg.

In isolated atoms of alkali metals, which have the s^1 configuration of the valence electrons, s^2 configurations are formed, whose stability, like that of $s^x p^y$ configurations, decreases with an increase of the principal quantum number of the valence electrons. Beginning with potassium there occurs a strong disturbance of s^2 configurations owing to transition of part of the electrons to a d state. Since the s^2 configurations are inert, the toxicity of the metals is due to the presence of unpaired s^1 electrons, which can either accept electrons on interacting with substances of the organism, forming s^2 configurations, or donate an s^1 electron. For lithium the most stable is the s^1 configuration, and the stability of the bond decreases sharply in sodium with an increase of the principal quantum number. This decrease ought to continue, but beginning with potassium some of the electrons (unpaired) depart for the d shell, causing an increase of toxicity rather than a decrease, which would occur in the case of maximum filling of the d levels in K, Rb, and Cs. Because of this, sodium is not an exception. The LD_{50} for Li^+ is 14.26, for K^+ 8.31, Rb^+ 9.50, and Cs^+ 8.67 mg-atom/kg; the LD_{50} for sodium is 44.52 mg-atom/kg.

Metals of the copper group have the electronic configuration $d^{10}s^1$ and can donate one electron; at the same time the metals of the zinc group (electronic configuration of the isolated atom $d^{10}s^2$) can yield electrons chiefly of the inert s^2 groups, which should lead to a decrease of their toxicity. However, as a consequence of the appearance in cadmium and mercury of vacant f levels, to which partial transfer of s electrons is possible with disturbance of s^2 groups, their toxicity should increase considerably; this applies most of all to mercury for which this transition should be manifested more strongly as a consequence of the greater stability of the f states. In contrast to this, the transfer of electrons of

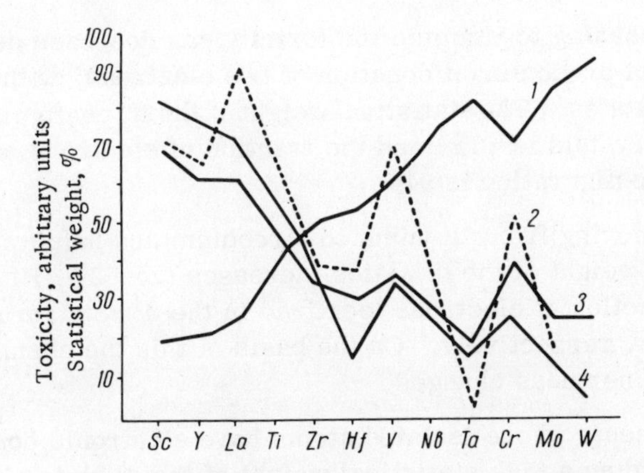

Fig. 38. Correlation between toxicity of transition metals and statistical weight of delocalized electrons in stable d^0 and d^5 configurations: 1) statistical weight of d^5 configurations; 3) of delocalized electrons; 4) of d^0 configurations; 2) toxicity of metals expressed in arbitrary units $\frac{LD_{50}Na = 44.5 \text{ mg-atom/kg}}{LD_{50}Me}$.

silver and gold to the vacant f levels should lead to a decrease of toxicity, especially if the simultaneous increase of the stability of the d configurations is taken into account.

Comparing the activity of the elements of this group, we find that copper and its compounds, although they have a toxic effect, affect the organism much more weakly than cadmium and especially mercury. Whereas the LD_{50} for Cu^{2+} is 0.45 mg-atom/kg, for Cd^{2+} it is 0.033, and for Hg^{2+} 0.0195 mg-atom/kg. The same regularity is noted on comparing the MAC, which for copper, according to our data, is equal to 0.5 and for mercury 0.001 mg/m^3.

In the light of what has been said above, the toxicity of transition metals with a d shell being completed decreases with an increase of the stability of the corresponding d configurations of electrons and their statistical weight (Fig. 38). It is necessary to take into consideration that the d^5 configuration is the most stable energetically, then the d^{10}, and the least stable is the d^0 configuration. Then we should expect from scandium (d^1s^2) mainly donation of electrons, with the appearance of stable d^0 configurations and relatively high toxicity. The statistical weight of the d^5 states in metallic crystals of scandium amounts to 18%, while the fraction of electrons localized in the d^5 configuration is 30% [312].

On passing to titanium the toxicity can decrease as a consequence of predominant donation of two electrons, with the formation of d^0 states. The statistical weight of the d^5 configurations in metallic crystals is 43%, and the fraction of electrons localized in the d^0 configuration is 94%.

On moving from titanium to zirconium and hafnium the statistical weight of the d^5 states increases (Zr 52%, Hf 55%), and the fraction of electrons localized in the d^0 configuration is 65 and 55%, respectively. On the basis of this the chemical and biological inertness changes.

Elements whose isolated atoms have electronic configurations providing a high statistical weight of the d^5 states in metallic crystals should have a relatively low toxicity: The higher the statistical weight of the d^5 states, the lower should be the toxicity. Indicative in this respect is the decrease of toxicity from vanadium (d^3s^2) to niobium (d^4s^1) and tantalum (d^3s^2) in connection with an increase of statistical weight (V 65%, Nb 76%, Ta 84%) and of energy stability of the d^5 configurations with increase of the principal quantum number. The toxicity decreases similarly on proceeding from chromium to molybdenum and to tungsten, the statistical weights of the d^5 states for which are respectively 73, 88, and 94%, and the percentage fractions of localized electrons are 58, 73, and 74.

The atoms of manganese and chromium, even in the isolated state, have respectively the d^5s^2 and d^5s^1 configurations of the valence electrons, i.e., a high statistical weight of the stable d^5 configurations. When the atoms of these elements become joined to form a crystal, the stable d^5 configurations are preserved completely, and the s electrons are practically free, since their participation in the formation of stable d configurations is impossible. This state of s electrons causes the high toxicity of Mn and Cr; in the case of formation of metallic Cr crystals from isolated atoms the s electrons can be joined partially into the stable s^2 configurations, whereas in the case of formation of crystalline manganese partial disturbance of the s^2 configurations is necessary for the occurrence of a chemical bond between atoms. Thus the relative concentration of existing free electrons should be higher for Mn than for Cr, which in a number of cases can cause metallic manganese to be more toxic than metallic chromium.

The metals of Group VIII (iron triad and platinum metals) are characterized by a high statistical weight of stable d^5 and d^{10} configurations (Fe 78 and 22%, Ni 3 and 97%, Co 32 and 68%) with little possibility of transfer of valence electrons to a delocalized state (Fe 23.7%, Co 5%, Ni 3%), which should determine their comparatively low toxicity. The latter increases with an increase of statistical weight of the d^{10} configurations and decreases with an increase of the principal quantum number, i.e., on progressing from metals of the iron triad to ruthenium–rhodium–palladium and to osmium–iridium–platinum. With a slight excess of electrons at the d levels (not more than five) the formation chiefly of d^0 configurations with transfer of the "excess" electrons is possible, and with the number of d electrons close to 10 the predominant

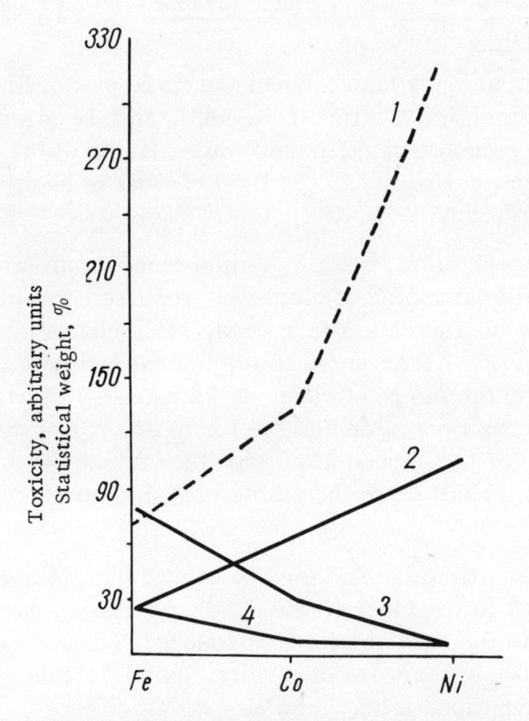

Fig. 39. Correlation between toxicity of elements of the iron group and statistical weight of d^5 and d^{10} configurations: 1) toxicity of metals, arbitrary units; 2) statistical weight of d^{10} configurations; 3) of d^5 configurations; 4) of delocalized electrons.

formation of d^{10} configurations with a decrease of the number of delocalized electrons is highly probable. Since the d^{10} configurations are less stable than the d^5 configurations, an increase of their statistical weight should lead to an increase of toxicity (from Fe to Ni, from Ru to Pd, from Os to Pt), which is in fact observed (Fig. 39).

Thus a comparison of the data on the toxicity of various elements with their electronic structure indicates a definite correlation between the biological activity of substances and the formation of stable electronic configurations of atoms. With an increase of the statistical weight of the latter due to localization of valence electrons and decrease of delocalized electrons, the toxicity decreases.

Regularities of the Toxic Effect of Transition Metal Oxides

Much is already known about the toxic properties of oxides of transition metals [88, 162, 168, 263], and the information, despite its incompleteness, permits discussion of the causes of the differences in toxicity in the light of what is known of their electronic structure [137, 187].

The oxygen atom, having the electronic configuration s^2p^4, is a very active acceptor of electrons required for building up its electronic configuration to a stable state characteristic of an inert gas (neon). At the same time, interaction with very strong acceptors creates the possibility (in rare cases) of partial donation of electrons by oxygen and the formation of the electronic configuration of the carbon atom s^2p^2 with further transformation through $s \rightarrow p$ transition to the stable sp^3 configuration characteristic of diamond.

The transition metals, tending to build up their electronic configurations to the most stable d^5, d^0, d^{10} states, during reaction with oxygen in turn act as electron donors. Their donor capacity decreases with an increase of localization of valence electrons to stable configurations from Sc to W.

This reaction of atoms of transition metals with oxygen results mainly in the formation of covalent bonds by localized electrons and covalently-bound groups of oxygen atoms.

To facilitate examination of the problems of interest to us, we can tentatively divide the transition metals into three groups according to their electronic structure.

The first group includes elements characterized by a pronounced tendency to donate electrons with the formation of the stable d^0 electronic configuration: Ti ($3d^2 4s^2$), Zr ($4d^2 5s^2$), and Hf ($5d^2 6s^2$).

The second group is composed of transition metals having less strong donor properties and forming the stable d^5 and d^{10} electronic configurations: V ($3d^3 4s^2$), Nb ($4d^4 5s^1$), Ta ($5d^3 6s^2$), Cr ($3d^5 4s^1$), Mo ($4d^5 5s^1$), W ($5d^4 6s^2$), Mn ($3d^5 4s^2$), Tc ($4d^5 5s^2$), and Re ($5d^5 6s^2$).

The third group is made up of elements with donor–acceptor properties and the d^{10} electronic configuration, i.e., elements of Group VIII of the periodic system: Fe ($3d^6 4s^2$), Co ($3d^7 4s^2$), Ni ($3d^8 4s^2$), Ru ($4d^7 5s^1$), Rh ($4d^8 5s^1$), Pd ($4d^{10} 5s^0$), Os ($5d^7 6s^1$), Ir ($5d^7 6s^2$), and Pt ($5d^2 6s^1$).

We will begin the presentation of data on the toxicity of oxides with elements of the second group, since it has been studied most completely for these oxides. The following median lethal doses of oxides of certain metals, expressed in terms of metal content, have been established in experiments involving administration to mice [263]:

Oxides	LD_{50}, mg-atom/kg	Oxides	LD_{50}, mg-atom/kg
BeO	82.5	TiO_2	165.6
MgO	61.6	VO_2	0.615
CaO	54.6	V_2O_5	0.180
BaO	1.95	Ta_2O_5	16.0
Al_2O_3	70.6	CrO_3	112.0
GeO_2	20.6	MoO_2	3.3*
PbO_2	1.62	MoO_3	0.97
PbO	0.975	MnO_3	3.9
CuO	3.41	CoO	3.14
ZnO	2.25	Co_2O_3	35.5
CdO	0.124	Ni_3O_4	4.0
HgO	0.046		

*Established by O. Ya. Mogilevskaya, 1950 (on rats).

Metals of this group form not only oxides of the general formula MeO but also higher oxides: MeO_2, Me_2O_5, MeO_3.

The most toxic of the vanadium oxides is V_2O_5 for which the LD_{50} established in experiments on white mice with oral administration is, expressed as vanadium content, 23.4 mg/kg or 0.18 mg-atom/kg.

For vanadium trioxide V_2O_3 the LD_{50} is 130.6 mg/m^3 [263]. Vanadium monoxide VO is also relatively nontoxic, its LD_{50} expressed as vanadium being 0.276 mg-atom/kg [394].

The main symptoms of acute and chronic poisoning by vanadium compounds are an irritating effect on the mucous membranes of the air passages and affection of the structure and functions of some internal organs (lungs, liver, kidneys). Vanadium pentoxide causes more pronounced pathological changes; under the action of vanadium tri- and dioxide such changes are much less marked [263].

The most toxic among the oxides of manganese are the higher oxides.

According to Levina [162], endotracheal administration of MnO_2 in a dose of 10 mg/kg killed 60% of experimental animals and Mn_2O_3 in a dose of 20 mg/kg killed 20%, while MnO was not lethal even in a dose of 50 mg/kg. This regularity was observed also in the case of subcutaneous injection of these substances into mice. The lethality of the animals varied following the subcutaneous injection of different specimens of MnO_2 in a dose of 250 mg. One of them killed 20% of the animals, all mice were killed by the second, and only 10% died after injection of the third. Manganese trioxide Mn_2O_3 killed 20% of the mice, whereas after the administration of 300 mg of a mixture of MnO and MnO_4 all the animals survived.

After the subcutaneous injection of MnO and MnO_2 into rats an increase of the weight coefficients of the internal organs (lungs, heart, liver, kidneys, spleen) was observed, especially after the injection of MnO_2.

A higher toxicity of higher oxides in comparison with that of lower oxides was noted among chromium compounds. These oxides are divided into three groups according to their level of toxicity. Hexavalent chromium compounds are the most toxic, the dichromates being more toxic than the chromates. Trivalent chromium compounds are less toxic, and metallic chromium and its divalent compounds belong to the category of low-toxicity substances [88, 161, 263].

Some authors consider metallic chromium and its chromic and chromous compounds to be completely harmless, whereas hexavalent chromium compounds are 100 times more toxic than trivalent, which is due to their solubility [68, 263].

The toxicity of the higher oxides of molybdenum [224], tungsten [201], niobium, and tantalum [97, 98] is higher than that of the lower oxides, and lower than that of manganese, chromium, and vanadium.

The toxicity of the oxides of the metals arbitrarily placed in the third group (Fe, Co, Ni, etc.) has been studied less in comparison with the toxicity of the oxides in the second group. The toxicity of the higher oxides of elements of this group is less than that of the lower. It is known that Fe_2O_3 is a low-toxicity compound having a caustic action, whereas FeO has a general toxic action. The administration of FeO caused paralysis and death of animals, which was accompanied by convulsions and other phenomena [161, 376]. The data of [123, 394, 567] indicate that the toxicity of Co_2O_3 is less than that of CoO and Co. Whereas a dose of 25-50 mg of Co_2O_3 did not kill animals, it corresponded to LD_{50} for CoO and was lethal for Co. The nickel oxide Ni_2O_3 did not cause poisoning in a dose of 50 mg, although it was more soluble than reduced nickel [226].

There are no data in the literature on the toxicity of ruthenium, rhodium, palladium, and their oxides. The effect of other elements of Group VIII on the organism has received only little study. According to foreign authors, osmium does not have a toxic effect, whereas osmium tetroxide OsO_4 irritates the air passages and mucous membranes, causing corneal opacity, bronchospasm, pneumonia, headaches, insomnia, inflammatory diseases of the kidneys, and dermatitides with phenomena of tissue necrosis. The toxic concentration of OsO_4 is 0.001 mg/m^3.

Platinum oxides have an irritating effect: PtO_2 causes fragility of the nails, eczema, allergic diseases; PtO_3 causes acute and chronic dermatitides. The maximum allowable concentration for platinum and its compounds, adopted in the USA, is equal to 0.002 mg/m^3 (expressed as the element) [161].

The administration of the dust of metallic titanium and TiO_2 induced in the lungs moderate proliferation of the connective tissue, considerably less than that effected by SiO_2. After endotracheal administration of other specimens of titanium dioxide, feeble parenchymatous degeneration of the liver and kidneys and a slight increase of collagen in the lung tissue were observed. The pathological changes occurring in the lungs were limited to a proliferative cell reaction at sites of dust localization and hyperplasia of the peribronchial lymph nodes with their subsequent slight sclerosis [220, 221].

After endotracheal administration of metallic zirconium and its dioxide changes occurred only in the lung tissue and were characterized by a proliferative cell reaction around the dust particles with the development of feeble sclerosis [225, 530].

During the first months after the endotracheal administration of hafnium oxide (HfO_2) only a feeble pneumoconiotic reaction occurred in the lung tissue, and after nine months moderate fibrosis and mild phenomena of parenchymatous degeneration of the liver developed [105].

Thus oxides of the three groups of transition metals differ in toxicity. The biological activity of oxides of the first group of elements with $nd < 5$ (Ti, Zr, Hf) is low. Their toxic effect differs only negligibly from the effect of the metals. The formation of higher oxides is not characteristic of this group.

The higher oxides of the transition metals of the second group, for which $nd \leq 5$ (V, Cr, Mn, Mo, W), are more toxic in comparison with the lower oxides and with the metals. The most toxic are the hexavalent (MeO_3) and heptavalent oxides, whereas the biological activity of the lower oxides (MeO) resembles that of the metallic components, although the oxides can nevertheless be shown to have a more pronounced pathological effect.

In contrast to this, in the third group, for which $nd > 5$ (Fe, Co, Ni, etc.), the lower oxides exhibit a higher activity in comparison with higher oxides. On the whole the toxicity of oxides duplicates the aforementioned regularity in the effect of transition metals, which depends on the possibility of their forming stable electronic configurations, and is also linked with a change of the energy stability in relation to the value of the principal quantum number.

These differences in the biological effect of the higher and lower oxides and of metals can be explained by the capacity for building up the electron shells during formation of oxides to energetically stable electronic states. From this viewpoint we can imagine that when atoms of transition metals of the first group react with oxygen, the transfer of one or two d electrons for building up the s^2p^4 shell oxygen to the stable s^2p^6 electronic configuration of inert gas probably occurs easily. In this case a high statistical weight of the stable d^0 configurations of the metal atom and the s^2p^6 configurations characteristic of inert gases is achieved.

The capacity of transfer electrons is less for atoms of the transition metals of the second group than for transition metals of the first group, and therefore during reaction with oxygen complete build-up of the s^2p^4 configuration to the stable s^2p^6 state does not occur. In this case there occurs a deficit in the equilibrium of the electronic structure of the outer shells of the metals, which is replenished through disturbance of electron exchange during oxidation-reduction reactions in the body fluids. In this connection, the oxides of these metals, particularly the higher oxides, induce pronounced fatty degeneration and cloudy swelling of the parenchymatous organs (heart, liver, kidneys), the activity of oxidative enzymes decreases, and oxidation of cysteine and other phenomena occur. It is known that the majority of metals interact with the body's biological media by blocking the active SH groups of the protein molecules. Gordi [73], using the paramagnetic resonance method, showed that loss of an electron of a sulfur atom is the first stage of inactivation of thiol-containing enzymes.

Both higher and lower oxides are formed when elements of the third group (Fe, Co, Ni, etc.), for which $nd > 5$, react with oxygen. However, in this case the opposite direction of the biological effect is observed. The lower oxides have a more toxic effect. Apparently the cause of this is the high statistical weight of the delocalized electrons of the lower oxides (FeO, CoO, NiO).

The iron atoms in an oxide tend to acquire a stable d^5 configuration, and a considerable proportion of the valence electrons after partial filling of the electron shells of oxygen to s^2p^6 pass into a collective state. CoO and NiO are mainly characterized by a high statistical weight of the d^{10} states, and hence, of the sp^3 configurations of the valence electrons of the oxygen atom. In this case the statistical weight of the d^{10} configurations increases from CoO to NiO, which is also responsible for the difference in their toxicity, i.e., a higher activity of NiO in comparison with the activity of CoO.

During the formation of higher oxides the statistical weight of the stable d^{10} configurations of the metal atoms and of the sp^3 and s^2p^6 configurations of the electrons of the oxygen atoms increases.

On the basis of the foregoing we can conclude that at present there are no complete data on the toxicity of oxides of various metals however, systematization of the material in the literature in relation to the electronic structure of metals and oxygen and conditions of formation of stable electronic configurations permits predicting a possible toxic effect.

The toxicity of metallic oxides in general obeys the regularities characteristic of simple substances: The level of activity decreases with an increase of the statistical weight of stable electronic configurations and decrease of the number of electrons in a collective state.

At the same time, when considering the possibility of formation of stable configurations in compounds it is necessary to take into account the original state of the metals, since depending on their form (α or β), their state will be more or less stable [137], and this in turn will affect the conditions of formation of oxides and their toxicity.

Dependence of the Toxic Effect of Compounds of Metals with Nonmetals on Their Electronic Structure

During the formation of a crystalline solid the interacting atoms, tending to acquire an energetically most stable state, donate or accept for their electron shells some valence electrons. The latter, building up the electron orbits, pass into a localized state. A part of the electrons is in a delocalized state, accomplishing a bond between stable configurations through electron exchanges.

The localized valence electrons form a broad spectrum of electronic configurations differing in their energy stability [308].

We can conclude from the foregoing that the metallic bond in compounds is due to the delocalized part of the valence electrons, while the covalent bond is due to the maximum statistical weight of the most stable s^2p^6 configurations.

Refractory compounds can be divided into two groups according to the type of chemical bond. Compounds formed upon reaction of nonmetals (boron carbide, silicon carbide, boron nitride, silicon nitride, and boron carbonitride) have a strong covalent bond. A metallic type of bond is characteristic of other compounds obtained by the reaction of transition metals with nonmetals [295].

The pathological changes induced in the organism by refractory compounds of nonmetals are characterized chiefly by a productive process in the lung tissue and bronchi with subsequent feeble sclerosis; in addition, these changes were found to be accompanied by collagen formation. After administration of boron and silicon nitrides the maximum increment of collagen in the lung tissue was 46.6-50.2% and after administration of boron and silicon carbides, 76-120% (in comparison with the control).

In addition to the common features of pathological disorders induced by refractory compounds, there are also peculiarities specific for each group. Materials belonging to the boride group must be considered the most active of all compounds.

Of the six metal borides investigated (CaB_6, Mo_2B_5, Cr_3B_2, TiB_2, ZrB_2, NbB_2), CaB_6 has the most active toxic effect, inducing acute and chronic poisoning of animals, moderate anemia, disturbances of nitrogen metabolism, and severe changes in the structure of internal organs. Ranking second in toxicity is Mo_2B_5, which was found to induce acute and chronic poisoning; the LD_{50} for this compound was slightly lower than for CaB_6.

The remaining borides did not kill animals, but as a result of their action on the organism degenerative changes of varying degrees occurred in the parenchymatous organs, as well as pulmonary disorders and other shifts.

Thus the borides can be arranged in the following decreasing order of expressivity of toxic phenomena: borides of calcium, molybdenum, chromium, titanium, zirconium, and niobium.

Unlike the borides, metallic carbides (chromium, titanium, and zirconium carbides) caused parenchymatous degeneration in the liver and kidneys, and feeble productive processes in the lung tissue. Here, as in the case of the borides, the expressivity of the pathological disturbances decreased from chromium carbide to titanium and zirconium carbides.

Shortly after the administration of boron carbide only a feeble fibrogenic effect and an inflammatory-proliferative reaction were noted in the lung tissue and bronchi.

During the formation of borides of transition metals, the boron atoms, having in an isolated state the s^2p configuration and tending to acquire through $s \rightarrow p$ transition the more stable sp^2 configuration and then sp^3, form the latter as a result of electron exchange between boron atoms according to the scheme $2sp^2 \rightarrow sp^3 + 2e$. This is responsible for the formation in the boride lattices of structurally isolated electron pairs, chains, networks, and skeletons [291]. As a result of such regrouping some of the valence electrons are released, and these become delocalized and can then participate in the development of pathological processes.

The same occurs also in amorphous boron and boron carbide, for which capture of electrons with an increase of the statistical weight of the sp^3 configurations and donation of electrons, which pass into a delocalized state as a consequence of energy regroupings of the boron atom, are possible. During the formation of boron carbide some of the delocalized electrons, released during exchange between the boron atoms, can be expended in stabilizing the sp^3 configuration of the carbon atoms, and consequently the biological activity of boron carbide is lower than that of boron. This is confirmed, for example, by the decrease in the collagen content.

Upon reaction of calcium (s^2) and boron (s^2p) atoms the s electrons of the metals, being excited, can partially pass to the completely vacant d level. In turn the boron atoms, as a consequence of the regroupings indicated above, can release electrons, a part of which goes for building up the s level of the metal, while a considerable part of them remains delocalized and is responsible for the toxic effect.

During the formation of Mo_2B_5, as a result of the reaction of Mo (d^5s^1) and B (s^2p) atoms, the s electrons of the two metal atoms can be used for building up the sp^3 configuration of only one boron atom. The other four atoms would be expected to ensure a stable state through internal regrouping of the electrons by the scheme indicated, a considerable part of them becoming delocalized.

Weakening of the toxic effect on progressing from chromium compounds to titanium and zirconium compounds is related with the fact that for chromium (d^5s^1) the valence shells can be built up by atoms of nonmetals, which is accompanied by disturbance of their stable configurations and delocalization of electrons. For titanium and zirconium, having a small statistical weight of atoms with stable configurations, there occurs mainly the transfer of delocalized electrons to atoms of the nonmetal with the formation of stable configurations, which makes them less active in reactions involving the manifestation of a toxic effect.

During the formation of transition metal carbides (Cr_3C_2, TiC, ZrC, HfC) some of the delocalized electrons of the metal atoms pass to the carbon atom, stabilizing the sp^3 hybrid configurations of the latter [306, 307]. An isolated carbon atom has the s^2p^2 configuration, which can be rearranged to sp^3 through $s \rightarrow p$ transition, but this occurs only under special conditions, for example, in the formation of diamond. Under ordinary conditions a set of the s^2p^2, sp, sp^2, sp^3 configurations is formed, which are characterized by a tendency to change to the energetically most stable sp^3 configuration through attraction of delocalized electrons of partners (e.g., transition metals).

Thus the electronic structure of the carbides does not differ fundamentally from the electronic structure of the corresponding transition metals, if we disregard the slight increase of the statistical weight of the sp^3 configurations of the carbon atom and decrease of the fraction of delocalized electrons of the metals. Therefore, the toxicity of carbides is close to that of the metals.

A study of the biological effect of the dust of refractory compounds of the silicide group ($TiSi_2$, $MoSi_2$, WSi_2) shows that under their action the content of collagen and ascorbic acid in the lung tissue increases, nonspecific interstitial bronchopneumonia with cellular-dust nodules develops, peribronchial and perivascular infitrates occur; under the action of molybdenum disilicide perivascular edema develops. Degenerative changes in the liver and renal convoluted tubules are also noted. Molybdenum disilicide increases the permeability of the capillaries of the renal glomeruli, as a consequence of which an albuminous exudate accumulates in the cavity of the capsules.

A comparison of the data on the pathological effect of the substances investigated with data on the toxicity of the metals in these substances shows that the effect of the silicides of titanium, molybdenum, and tungsten, corresponding in general to the toxicity of their metallic components, differs only by an increased fibrogenicity.

During the formation of transition metal silicides some of the delocalized electrons of the metals transfer to the silicon atoms with an increase of the statistical weight of the stable sp^3 configurations characteristic of silicon. The moderate chemical activity of elemental silicon and its low toxicity are related with the occurrence of these configurations. If the statistical weight of the sp^3 configurations of the silicon atoms is sufficiently high, the chemical activity and toxic effect of the compounds should not differ substantially from those for the metals forming these compounds.

The aforesaid pertains fully to the formation of titanium disilicide. In the case of the silicides of molybdenum and tungsten, however, whose metallic components have a high statistical weight of the d^5 configurations and accordingly a low content of delocalized electrons, the stabilization of the sp^3 configurations is accomplished mainly by electron exchange between the silicon atoms. Ultimately the electronic structure of all three silicides being considered becomes approximately the same, and it is this that accounts for their similar toxicity and the similarity of the toxic properties of the silicides and corresponding metals, especially in the case of molybdenum and tungsten.

The changes induced in the organism by boron nitride BN, silicon nitride SiN, and boron carbonitride BNC, having strong covalent bonds, are limited mainly to feeble pneumoconiotic disturbances and inflammatory-proliferative reactions in the bronchi, whereas in the case of metallic nitrides (TiN, ZrN, AlN) the pathological alterations were found to be characterized by a feeble vascular reaction (swelling of the walls of small- and medium-bore vessels) and by degenerative changes in the parenchymatous organs.

After endotracheal administration of niobium nitride changes in the form of cellular-dust reactions and emphysema were observed only in the lung tissue. Pathological disorders were absent in other organs. Such changes were less pronounced than those induced by niobium pentoxide and other niobium compounds [97].

Taking into account the electronic structure of nitrides, we must first of all divide them into two groups: nitrides of nonmetals (boron, silicon, aluminum) and nitrides of transition metals (titanium, zirconium, niobium).

During the formation of the nitride phases of nonmetals there occur strong hybrid localizations of the valence electrons of the cores of the nitrogen and nonmetal atoms, which provides these compounds primarily with a covalent type of bond. Thus, during the formation of boron and silicon nitrides there is chiefly an increase of the energy stability of the valence sp configurations owing to s → p transition with the transformation of the s^2p electronic configuration of the isolated boron atom to sp^2. At the same time the nitrogen atom acquires, as a consequence of s → p transition, a sp^4 configuration and then readily gives up an electron, resulting in transformation to a quasi-stable sp^3 configuration identical to the electronic configuration of the carbon atom in diamond. The electron released in this case transfers to the nonmetal atom, with the formation of a sp^3 configuration ($p^2 + e = p^3$).

Metals reacting with nitrogen form a stable d^5 configuration; the nitrogen atoms form sp^3 and s^2p^6 configurations depending on the character of the electronic structure of the metal.

According to calculations made on the basis of x-ray spectral investigations, in crystals of titanium, zirconium, and niobium the statistical weight of atoms having the stable d^5 configurations of localized electrons is respectively 43, 52, and 76% [312]. During the formation of nitrides of titanium and zirconium, which have a high fraction of delocalized electrons, acceptance of electrons from the nitrogen atoms is less probable than transfer of electrons to nitrogen, with the accompanying formation of relatively high statistical weights both of the d^5 configurations of the metal atoms and of the s^2p^6 configurations of the nitrogen atoms. The latter also provide a sufficiently high chemical inertness and a lower pathological activity of titanium and zirconium nitrides than those of the corresponding metals. The statistical weight of the d^5 configurations of niobium is sufficiently high already in the elemental state, and therefore during the formation of niobium nitride we can expect the occurrence of sp^3 configurations in addition to the occurrence of s^2p^6 configurations of the nitrogen atoms as a consequence of the attraction of delocalized niobium electrons.

In the case considered the statistical weight of stable electronic configurations will be high, and their energetic stability would be expected to be higher than for titanium and zirconium nitrides. As a consequence of this the pathogenic activity of niobium nitride is less than the activity of titanium and zirconium nitrides. Another process can occur simultaneously, viz., the nitrogen atoms can acquire the valence electrons of boron or aluminum with the formation of the stable s^2p^6 configuration by both partners (for nitrogen $s^2p^3 + 3e = s^2p^6$).

The formation of a high statistical weight of the sp^3 configurations in the compound of boron with nitrogen leads to the formation of a cubic (diamond-like) modification of boron nitride – so-called borazon. In the case of a high statistical weight of the s^2p^6 configurations hexagonal structures of ordinary boron and aluminum nitrides are formed. Obviously, the electronic structure of nitrides is in fact intermediate between the cases considered, i.e., there is a set both of sp^3 and s^2p^6 configurations of localized electrons with varying statistical weights. However, in all cases the valence electrons are localized chiefly into stable configurations, and the portion of delocalized electrons is small, which gives rise to dielectric properties, considerable width of the forbidden band, low thermal conductivity, and high chemical inertness in nitrides of nonmetals of Group III. Thus, the weak biological activity of boron nitride and carbonitride and the decrease of toxicity of aluminum nitride is due to the high statistical weight and high energetic stability of the electronic configurations of their atoms, which have a low statistical weight of delocalized electrons responsible for the pathogenic activity of various elements and the given nitrides.

Analogous arguments pertain to silicon nitride, whose lattice is characterized by a high statistical weight of stable electronic configurations of the silicon and nitrogen atoms. For the same reasons the toxicity of boron nitride approaches the toxicity of boron, whose chemical inertness is due to a high statistical weight of energetically very stable electronic configurations. It should be noted that there are no substantial differences in the electronic structure of these nitrides. A slight increase of toxicity (boron nitride $<$ silicon nitride $<$ aluminum nitride) due to a decrease of the energetic stability of the sp configurations is observed. The latter should play a particularly important role at high temperatures. In this connection, for work with these substances in hot shops we must recommend a slight decrease of their maximum allowable concentration; in accordance with the aforesaid this decrease should be a maximum for aluminum nitride, less for silicon nitride, and a minimum, if any, for boron nitride.

Thus the results of examining the biological activity of various groups of refractory compounds permit the conclusion that the biological effect of compounds, like that of simple substances, depends on the statistical weight of stable electronic configurations and energetic stability in their principal components. The difference in the toxic properties of compounds belonging to the chalcogenide group can be explained by the formation of stable electronic configurations.

According to current concepts [311], the isolated atom of a chalcogen has the s^2p^4 electronic configuration and in compounds tends to capture electrons of partners with building up of the outermost shell to the most stable s^2p^6 configuration. However, under certain conditions there exists the probability of the chalcogen atom giving up a part of the electrons with the formation of a quasistable sp^3 configuration.

A comparison of the data on melting points and other physicochemical properties [190, 300, 301] established that among the chalcogens the energetic stability decreases with an increase of the principal quantum number on progessing from sulfur to selenium. For selenium the statistical weight of delocalized electrons increases and exchange between the electrons thus released and the sp configurations is activated.

For tellurium, in connection with the presence of a vacant 4f level, a part of the delocalized sp electrons passes to the unfilled f level, whose large quantum number and energetic stability lead to increased formation of stable electronic configurations, in connection with which the melting point also rises. Under such conditions the sp configurations are also disturbed, as a consequence of which the statistical weight of the delocalized electrons increases. The appearance in tellurium of metallic properties, which are manifested in its reactions with other elements, is to some extent related with this.

In the case of the formation of chalcogenides of alkali metals, for example, sodium sulfides, the s^2p^4 configurations of sulfur are built up to the stable state of the inert gas argon, through the transfer of one electron of the s shell of the sodium atom. In this case compounds such as Na_2S and other polysulfides are formed. It has been established that in the group of alkali metals the ability to form sulfides increases with an increase of the ionization potential of the metal atoms.

During the formation of sulfides the heat of formation per sulfur atom decreases, which had been attributed to the appearance of the stable s^2p^6 configurations upon transfer of two s electrons to the sulfur atom. An increase of the number of sulfur atoms in sulfides results in the formation of covalent bonds between them, as a consequence of which the crystal lattice is loosened. This causes a decrease of the strength of the internal bonds, a lowering of the melting point, and a change of other physical characteristics, and also leads to an increase of toxicity (e.g., for sodium and potassium sulfides).

When transition metals react with chalcogen atoms the d-electron shell is built up to the stable d^5 and d^{10} configurations with the formation of chalcogenides of various compositions: Me_2Ch, MeCh, Me_3Ch, and others; sulfides, selenides, and tellurides are usually isostructural and in most cases are compounds of variable composition. For small Ch/Me ratios the metallic bond, realized by collective electrons, predominates in chalcogenide compounds. As the quantity of chalcogens increases, particularly sulfur atoms, the share of the covalent bond increases [311].

As a result of examining the toxic properties of the sulfides investigated, of molybdenum MoS_2 and tungsten WS_2, and of the selenides $MoSe_2$, WSe_2, and $NbSe_2$, we can assume that the relative inertness of molybdenum and tungsten sulfides observed in our experiments is related with the occurrence of a high statistical weight of stable electronic configurations as a consequence of electron exchange between the molybdenum (d^5s^1) and sulfur (s^2p^4) atoms in the former case and tungsten (d^4s^2) and sulfur (s^2p^4) atoms in the latter. Here stable d^5 configurations form as a result of the molybdenum atoms giving up valence s electrons, and as a consequence of the transition of electrons to the p shell of the sulfur atoms the latter is built up to the stable s^2p^6 configuration. Consequently, during the reaction of a molybdenum atom with a sulfur atom a high statistical weight both of the d^5 configurations of the metal atom and of the stable s^2p^6 configurations of the sulfur atom is attained, which is responsible for the high inertness, semiconducting properties, refractoriness, etc. of MoS_2.

When the d^4s^2 atoms of tungsten react with the s^2p^4 atoms of sulfur acceptance of tungsten electrons by sulfur atoms is also probable, which is related with the tendency of the sulfur atoms to form the more stable s^2p^6 configuration.

Thus, in the case of the formation of WS_2 a high statistical weight of the d^5 configurations of the metal atoms and stable s^2p^6 states of the sulfur atoms is secured, which determines the inertness of this compound.

The increase of toxicity in the series of selenides $MoSe_2$, WSe_2, $NbSe_2$ is apparently related with a decrease of energy stability and increase of electron exchange between the released electrons and sp configurations of the selenium atoms. As a consequence of this the Se atoms are characterized by a definite statistical weight of the delocalized electrons.

The stable d^5 and s^2p^6 configurations are formed when molybdenum atoms react with selenium atoms. Some of the electrons are in a collective state, since there is already a definite statistical weight of the delocalized electrons in the selenium atom itself as a result of a decrease of its energetic stability with increase of the principal quantum number.

The interaction of tungsten atoms with delocalized selenium electrons is even more restricted, since tungsten atoms have a high energetic stability. This also determines the high content of delocalized electrons in tungsten diselenide and therefore the higher toxicity of the latter in comparison with the toxicity of molybdenum diselenide.

Upon interaction of niobium (d^4s^1), which is at a higher energy level than tungsten, and selenium (s^2p^4) the donor abilities of the metal increase. As a consequence of this the possibility of acceptance and stabilization of collective electrons of selenium increases in the formation of niobium diselenide $NbSe_2$. This apparently explains why niobium diselenide is less toxic than tungsten diselenide and more toxic than molybdenum diselenide.

On the basis of the foregoing we can assume that the tellurides, too, have a toxic effect, which depends on the conditions of reaction between the atoms of the metal and chalcogen, and also the ability to form stable electronic configurations in compounds. This was confirmed in our experiments.

These data on the toxicity of various refractory compounds and chalcogenides in comparison with their electronic structure permit the conclusion that the toxicity of compounds is often lower than the toxicity of the starting components, but its level in each specific case is determined by the conditions of formation of stable electronic configurations. A decrease of biological activity is associated with a decrease of the fraction of delocalized electrons in a compound. This is due to the donor abilities of the metals in connection with their tendency to form the energetically most stable d^5, d^0, and d^{10} configurations and to the ability of the nonmetal atoms to acquire delocalized electrons. In this case the energetically stable s^2p^6 configurations characteristic of inert gases, or in the event of partial donation of electrons, the less stable sp^3 configurations are formed.

An analysis of the electronic structure of various compounds apparently offers the opportunity to predict the toxic effect of various substances and hence to anticipate the need for various preventive measures aimed at eliminating the harmful effect of such compounds on workers.

It must be pointed out that the biological activity of compounds, like that of simple substances, correlates with the level of the melting point, chemical stability, and other physicochemical properties. This in turn is governed by the strength of the internal bonds, which ensure a particular crystal structure. Therefore, when evaluating the possible toxic effect of metallic compounds, we can use as auxiliary guides, in addition to the electronic structure, various physical and chemical constants and character of the crystal structure, since a higher melting point, hardness, and high crystal lattice symmetry (bcc structure) correspond to a high statistical weight of stable electronic configurations and therefore lower toxicity.

In the case of prolonged action on the organism of the dust of metals with a bcc structure (β-Ti, β-Zr, β-Hf, V, Nb, Cr, Mo, W, α-Fe) the pathological changes are characterized mainly by the development of a moderate productive process in the lung tissue with subsequent feeble sclerosis.

A low statistical weight of stable electronic configurations, among which low-stability d^0 configurations predominate, corresponds to simple substances having a relatively low melting point, high conductivity, good ductility (Se, Y, La, α-Ti, α-Zr, α-Co, Zn, Cd, Hg), and less perfect lattice symmetry (hexagonal, rhombic, monoclinic, etc.). During their formation a considerable number of the electrons are in a delocalized state, in connection with which these metals have high reactivity and consequently a definite toxic effect. They have an acute and chronic action, causing affection of many systems of the organism and first and foremost of the parenchymatous organs (liver and kidneys).

An intermediate position is occupied by metals having a fcc structure formed by atoms with a high statistical weight of the d^{10} configurations (β-Co, Ni, Rh, Pd, Cu, Ag, etc.), which are characterized mainly by a chronic character of toxic action, with affection of the parenchymatous organs.

However, analysis based on peculiarities of the crystal lattice structure of substances does not enable us to take into account the fine differences in the toxic effect of metals related with a change of the energy stability of the atoms in relation to the principal quantum number. The formation of particular crystal structures is inherent to a group of metals united by certain energy characteristics of the structure of the electron shells. Therefore, on the basis of the crystal structure we can determine only approximately the character of the toxic effect and tentative maximum allowable concentrations of hazardous substances in the air of industrial rooms. Taking into consideration the toxic effect and established MAC's, as the approximate allowable concentrations we can recommended 6-10 mg/m^3 for substances with a bcc structure, 4-5 for fcc, 1-2 for cph, and for rhombohedral, monoclinic, and other low-symmetry structures 0.5 mg/m^3.

The formation of a bcc structure is characteristic also of low-toxicity nonmetals. As a consequence of the high statistical weight of stable sp^3 electronic configurations the carbon (diamond) and silicon atoms form a diamond-like structure, whereas graphite has a hexagonal layered lattice. The high toxicity of crystalline boron is related with monoclinic and rhombic structures.

A decrease of symmetry of the crystal lattice is characteristic also of the toxic elements Ga, Zn, Tl, Pb, As, Sb, Bi, Se, Te, and certain s metals.

In metallic compounds the crystal structures, as in the case of simple substances, are determined by the electronic structure and correlate with the character of pathogenic activity. Because of this, when evaluating the toxicity of new compounds it is logical to be guided by the degree of symmetry of the crystal structures formed.

Toxic Effect of Some Complex Compounds

Among complex compounds are the majority of vitally essential organic substances, including metalloenzymes [59]. Many metabolic processes are disturbed as a result of the formation of a complex of metals with proteins and nucleic acids. Antidote treatment of poisonings caused by various metals is closely related with the formation of stable complexes [101].

Owing to the complexity of these problems it is not possible to dwell in detail on them here, and therefore we will consider only the general relation between the toxic effect of some complex compounds, including metal carbonyls, and their electronic structure.

The literature data and the results of our investigations, presented in the preceding chapter, indicate a toxic effect of metal carbonyls. Its severity decreases from nickel tetracarbonyl $Ni(CO)_4$ to iron pentacarbonyl $Fe(CO)_5$ and further to cobalt tetracarbonyl $Co(CO)_4$, molybdenum hexacarbonyl $Mo(CO)_6$, and tungsten hexacarbonyl $W(CO)_6$.

Whereas the first two compounds, which are volatile liquids, cause acute and chronic poisoning, the other substances, having a crystal structure, cause only chronic poisoning.

The strength of the internal bonds in a complex is governed by the participation in their formation of the electrons of the central ion, which are at various energy levels. It was established that, other things being equal, bonds formed with the participation of the electronic cells of the p orbital are stronger than those involving the cells of the s orbital, while those formed with the participation of the d cells are stronger than the other two. Coordination bonds formed by electrons of all three orbitals simultaneously are the strongest.

Considering the structure of various metal carbonyls from this viewpoint, we can note that the strength of their coordination bonds is different. Since in the formation of carbonyls each group transfers two electrons, as many CO groups are added maximally to the metal atoms as is necessary for the electronic structure of the central atom of the metal to acquire the configuration of the inert gas following it.

In nickel carbonyl the metal atom has the electronic configuration d^8s^2, and for the formation of the stable configuration corresponding to the inert gas krypton ($d^{10}s^2p^6$) another eight electrons (d^2p^6) are needed. In this connection, in the formation of a complex compound it can accept only four CO groups, since each of them transfers two electrons to the central metal atom. Consequently, in nickel tetracarbonyl four coordination bonds occur through the participation of d^2p^6 electrons. In iron pentacarbonyl $Fe(CO)_5$ ten d^4p^6 electrons are lacking for building up the electronic d^6s^2 orbitals of the iron atom to a stable state. The coordination bond in this case occurs through the participation of a greater number of d electrons than in the preceding compound, as a consequence of which it should be stronger.

For cobalt d^7s^2 nine d^3p^6 electrons are lacking for the formation of the electronic configuration of inert gas, in connection with which, on forming a carbonyl compound, cobalt can acquire only four CO groups, and one electron remains in an unpaired state. However, since this state is energetically quite unstable, the cobalt carbonyl molecule assumes the paired structure $[Co(CO)_4]_2$.

Thus different numbers of coordination bonds appear in the formation of metal carbonyls belonging to the iron triad: in Fe, five bonds, two of them due to 4d electrons; in Co, four, including 1.5 due to the participation of the d orbital; in Ni, also four, one of which is due to the d-electron shell. Consequently, iron pentacarbonyl should have the strongest bond, then cobalt tetracarbonyl, and nickel carbonyl should have the weakest bond. However, the aggregate state of these substances and their toxic properties do not obey such a regularity. Iron and nickel carbonyls are liquids, which corresponds to a less stable state, and cobalt carbonyl is a solid crystalline substance. The toxicity of iron carbonyl, having two d bonds more than nickel carbonyl, is half that of the latter compound, for which only one coordination bond is formed due to the d cell.

Compared with the other compounds, cobalt carbonyl has a much lower toxicity. This is easily explained if we imagine that cobalt carbonyl, tending to achieve complete saturation of the electronic d cells, forms a paired structure. The crystalline state of cobalt tetracarbonyl and its lower toxicity are apparently related with this circumstance.

Still more d cells participate in the formation of the structure of molybdenum and tungsten hexacarbonyls. The d^5s^1 atom of molybdenum lacks 12 electrons ($d^5p^6s^1$) for building up its electronic configuration to a stable state.

Consequently, when these atoms react with CO groups six coordination bonds are formed as a result of the participation of the electronic cells of the three orbitals. In this case the d orbital can participate in the formation of two coordination bonds and transfer still another electron for the formation of a third bond.

In the formation of tungsten hexacarbonyl the d^4s^2 tungsten atom, tending to build up its configuration to the energy state of inert gas, can acquire also 12 electrons and form six bonds with six carbonyl groups, including three due to the d cells. Thus, comparing the possibility of formation of the strongest coordination bonds due to participation of d cells in both compounds, we can conclude that tungsten hexacarbonyl should be more stable than molybdenum carbonyl and consequently less toxic, which has in fact been established.

The more stable state of tungsten hexacarbonyl in comparison with the analogous molybdenum compound can be due to the high energetic stability of the tungsten atom as a result of an increase of the principal quantum number.

An examination of the toxic properties of various metal carbonyls in relation to the electronic structure of the starting complexing agents and primarily with the electronic configuration of the central atom indicates the existence of a definite correlation between the electronic structure of the compound, condition of formation of strong coordination bonds, and their toxicity. Knowing the electronic structure, we can predict the toxicity of complex compounds and consequently the hazard they constitute to workers' health. For example, chromium carbonyl $Cr(CO)_6$ also lacks 12 electrons ($d^5s^1p^6$) for building up the d^5s^1 configuration of the central chromium atom to the stable state corresponding to the inert gas krypton ($d^{10}s^2p^6$). In this case, just as for the molybdenum atom, two coordination bonds are formed due to the d cells, and the others are formed as a result of less strong interaction with the s and p shells.

It is necessary to take into account that the chromium atom has the least energetic stability in comparison with the atoms of molybdenum and tungsten. On the basis of this we can expect that chromium hexacarbonyl will be a less stable chemical compound and more toxic than the analogous compounds of molybdenum and tungsten.

Taking into account the similarity in the electronic structure of various complex compounds and proceeding from the data examined concerning the toxic properties of carbonyls, we can assume that the same regularity is retained among other complex substances as for the carbonyls. When evaluating the possible toxic effect of complex compounds it is necessary to take into consideration not only the electronic structure of the central ion but also the ligand. This is indicated in particular in [59], where an examination is made of the complexing of metals with proteins of the living organism (in metalloenzymes).

Metals form coordination bonds with active groups of protein molecules and thus markedly change the structure and properties of the latter. The active groups of proteins, arranging themselves around the metal ion, influence it. The symmetry of the arrangement of the functional groups and the force of the coordination bond change the properties of individual cations considerably. For example, iron in various proteins displays different properties depending on the active groups with which it is bound: with carboxyl groups (catalase), amino groups (peroxidase), or with the imidazole residues of protein (hemoglobin).

Each metal ion has specific properties which are manifested in preferential interaction with certain active groups [585].

Polarizability of the ligand and cations are of particular significance for preferential interaction and formation of a strong coordination bond in metalloenzymes and, apparently, in other protein complexes with metals.

It was found that the active groups of proteins, which are ligands in the complex, have different polarizability depending on the size of the ionic radius of sulfur, nitrogen, or carbon, The SH groups are the most capable of polarization, since the sulfur atom has the greatest radius; the nitrogen-containing groups (NH_3) have this property to a lesser degree [59], and the oxygen-containing groups (COOH) are the least polarizable, i.e., the most stable with respect to the electrostatic influence of the central ion in the complex compound.

The force of the polarizing action of the central ion is determined by its radius. The polarizability of the ion and its polarizing effect on surrounding ions, i.e., on active groups, decreases with an increase of ionic radius. In connection with this, such ions interact more actively with readily polarizable active SH groups of protein and less so with stable carboxyl groups.

Since the ionic radius is determined by the electronic structure of the atom, we can assume that ions of metals of the first large period, having the smallest ionic radius, possess the highest polarizability with respect both to the SH groups and to the amino and carboxyl groups of protein molecules. As a consequence of this, on interacting with the biological media they are most able to disturb the structure of protein molecules and, evidently, of nucleic acids, which apparently explains the high activity of transition metals of the fourth period.

Since the ions of these metals have a small radius and the sulfur ions have a large radius, steric hindrances to the formation of complexes of these elements with the SH groups of proteins are created. Hence we must consider that the formation of complexes with amino and carboxyl groups is more characteristic of transition metals of the fourth period.

As the atomic and ionic radius increases, the polarizability of the metal atoms decreases and therefore strong coordination bonds with protein structures can form only as a consequence of interaction with readily polarizable active groups; in connection with the aforesaid, the formation of coordination bonds with readily polarizable amino groups and particularly with sulfhydryl groups is more characteristic of transition metals of the fifth period. Ions of metals of the sixth period react weakly with protein molecules, mainly through coordination of the SH groups only.

On the basis of the results of an investigation into the strength of metal complexes in molecules of enzymes and nucleic acids it is concluded in [59] that the biological system competes successfully for a metal only if it can make available for binding geometrically analogous centers with a high binding force, i.e., readily polarizable anions or active groups.

In addition to polarizability, the stability of a complex, as noted in [59], is closely related with the character of the geometric structure of the complex and its coordination number. The ions of some metals form complexes in the form of an octahedral structure, others (Hg^{2+}, Cd^{2+}, Pb^{2+}, Cu^{2+}, Zn^{2+}) have a tetrahedral lattice with a small coordination number and form complexes with the sulfhydryl or amino groups of proteins. A third group of ions generally forms complexes with the nitrogen and oxygen-containing groups of proteins and has an octahedral lattice or another structure with a small coordination number. In this case the stability of the complex decreases in the following order: $Cu^{2+} > Ni^{2+} > Co^{2+} > Fe^{2+} > Co^{3+} > Cr^{3+} > Fe^{3+}$.

On the basis of examination of numerous facts it was established that the formation of a particular geometric structure of a complex is closely related with the valence state of the central ion, and also with the participation of particular active groups in the formation of the complex. Preferential action is attained as a result of limiting the number of free coordination bonds that can participate in the formation of the complex with the substrate.

The maximum specificity of metalloenzymes is related with the maximum saturation of the coordination bonds of the metals.

Relation between the Formation of Metallic Aerosols and Their Electronic and Crystal Structures

The electronic and crystal structures of metals and their compounds determines a number of properties: specific weight, cleavage, hardness, ductility, etc. The formation of condensation and disintegration aerosols of metals depends on these properties.

Of particular importance for the formation of disintegration (pulverization) aerosols is the crystallochemical characteristic cleavage, i.e., the ability of crystals to split along planes parallel to crystal faces under the action of mechanical forces [140]. Cleavage surfaces always correspond to faces with small indices. According to the ease of fracturing a crystal and the character of the cleavage surfaces of the fragments (smooth, interrupted, conchoidal, jagged) the following degrees of cleavage are distinguished: extra perfect (various micas, graphite), perfect, distinct, and indistinct (quartz, the majority of metals).

Crystals with indistinct cleavage are distinguished by a characteristic fracture: uneven, granular, conchoidal, and jagged. Polymorphous particles with sharp edges are usually formed upon comminution of such substances. Depending on their symmetry crystals can split in one direction (mica, gypsum, graphite), in two, in three, and in more than three (fluorite).

Cleavage is inseparably related with the characteristics of the crystal structure. Upon splitting, the crystal cleaves along certain planar networks of its structure, usually perpendicular to the direction of the weakest bonds. In the structure of graphite weak van der Waals bonds act perpendicular to the layers, which make possible easy separation of the planar networks from one another, and therefore upon disintegration of graphite dust particles of a lamellar shape are usually formed. This is also inherent in all structures characterized by distinct lamination.

In crystals with a chain structure cleavage as a rule is parallel to the direction of the axis of the chains. Cleavage along the faces of a prism is characteristic of tellurium crystals, in which the bonds between atoms or in spirallike chains are covalent, and between chains are van der Waalsian.

In heterodesmic structures cleavage also corresponds to the weaker bonds – van der Waals (graphite, talc), weak ionic (mica), weakened metallic (zinc, cadmium), or mixed.

The crystal structure and strength of the internal bond determines not only the shape of the dust particles but also their size, since mechanical disintegration of crystal structures occurs much more easily in substances having weak internal bonds. This promotes the formation of a more finely divided disintegration aerosol in comparison with an aerosol formed as a result of comminution of crystal structures with close cubic packing (for example, for chromium, tungsten, molybdenum, etc.).

When considering the toxicity of dust in relation to its particle size increased activity is usually linked with penetration of particles into deep respiratory passages. We suggest that the cause lies not only in this. With decrease in particle size the active surface of the particles increases markedly, the number of disturbed internal bonds increases, and as a consequence of this the reactivity of the particles, their solubility in body fluids, and the possibility of a transition to an ionic state and interaction with biopolymers (proteins, nucleic acids, etc.) increase.

The electronic structure of substances and the character of their crystal structure determine the possibility of fracture of the crystal structure under the action of mechanical forces. Many metals having a high statistical weight of stable electronic configurations and close cubic packing are, as was indicated, poorly amenable to mechanical disintegration. Upon formation of compounds with a predominance of the fraction of delocalized electrons the strength of the crystal lattice decreases and lessening of its structure occurs (for example, upon formation of oxides and especially higher oxides). This is the cause of the formation of dust of smaller particle size than in the case of the disintegration of the original metals. Such dust is therefore also more active, as indicated in [160, 280]. The same apparently occurs during the disintegration of materials such as interstitial solid solutions, etc. If the atomic radius of the nonmetal is greater than the radius of the interatomic space in the crystal lattice of the metal, loosening and weakening of the bonds in the crystal occur, which facilitates the disintegration of the material.

At present some investigators (138, 139, 373, 375, etc.) are attempting to establish a relation between the increase of activity of aerosols and the electrical charge of dust particles. We consider that this characteristic, too, is closely related with the electronegativity and donor–acceptor and other characteristics of elements determined by their electronic structure. The formation of electrical charges on the surface of dust particles can be explained by electrification of crystals either by a change of temperature (pyroelectricity) or by mechanical deformations such as compression (piezoelectricity).

The electrical charge of particles formed during disintegration can apparently be affected by the magnetic properties of the crystals, which in turn are related with the orbital movement of electrons around atomic nucleic and with their spin. In atoms of diamagnetic crystals the spins of individual electrons are mutually compensated. Crystals whose ions have the configuration of noble gases, or covalent crystals, belong to the diamagnetic class. Consequently, the presence of diamagnetic properties indicates covalent bonds in crystals.

Uncompensated electrons are present in atoms of paramagnetic and ferromagnetic crystals; the outer shells of these atoms are not filled.

The duration of suspension of dust particles in air is determined to a considerable extent by the specific weight of the parent substances of the particles. The specific weight depends on the atomic weights of the elements forming the crystal, ionic radii, valence of the ions, type of crystal structure, and "packing fraction." The larger the atomic weights of the elements, the higher the specific weight of the crystal. Light crystals consist of elements of the upper rows of the periodic system; the elements of the lower rows markedly increase the specific weight of crystals.

Irrespective of the size of the atoms, a smaller specific weight corresponds to a smaller packing fraction. The specific weight of crystals decreases when additional anions (OH, F, etc.) or water molecules are present in the structure, and increases with an increase in the coordination number.

Of considerable interest is the problem of the possibility of formation of condensation aerosols in various pyrometallurgical processes. Under high-temperature conditions the atoms of metals and other elements become excited, i.e., the spins are split with transition of electrons to higher energy levels. When metal vapors pass from a high-temperature zone into an environment of atmospheric oxygen (active electron acceptor) vigorous reaction occurs with the formation of oxides, which as a rule are found in the working zone. Since a mixture of metals with different melting points is subjected to thermal action, lower-melting-point compounds are released first (for example, manganese oxides).

The data examined in this chapter on the toxicity of simple substances, oxides, refractory and semiconducting materials, and certain complex compounds belonging to the carbonyl group in relation to their electronic structure indicate a close correlation between pathological activity and the character of the electronic structure of atoms and also the possibility of formation of stable electronic configurations in compounds. An increase of the statistical weight of stable electronic configurations reduces the pathogenic activity of a compound. As a result of the formation of strong coordination bonds between the central ion (complexing agent) and ligands, the toxicity of the complex compounds decreases.

It is also necessary to indicate the existence of a correlation between these characteristics of the electronic structure of substances and their capacity for disintegration, size of the dust particles formed during mechanical comminution of materials, and other characteristics of aerosols determining the possibility of development of pathological processes following their inhalation.

CHAPTER IX

Measures for Preventing Occupational Diseases among Workers in Powder Metallurgy Plants

A study of working conditions in powder metallurgy plants showed that the main production processes are accompanied by the liberation of appreciable quantities of dust, gases, and heat. In some shop areas the levels of noise and heat and gas liberation were excessive and an intensification of infrared radiation was noted.

Consequently, working conditions at powder metallurgy plants are characterized by a number of hazardous industrial factors which can be the cause of occupational diseases. This is confirmed by the results of investigations of the morbidity and state of health of workers and the data obtained in a study of the effects on the organism of various metal powders, refractory compounds, and other materials used most often as charge components and which are the source of industrial dust.

In connection with this, to prevent occupational diseases at powder metallurgy plants it is necessary to take a number of sanitary-engineering and technological measures aimed at preventing the release of dust into the air of the working zone and at eliminating the effect of other deleterious factors.

The content of metallic dust and gases in the air in various working zones should not exceed the maximum allowable concentrations laid down by sanitary laws and those recommended on the basis of investigating the toxic properties of metal powders, their mixtures, refractory compounds, chalcogenides, and metal carbonyls. The maximum allowable concentrations of various substances in the air of industrial rooms are given below.

Maximum Allowable Concentrations (mg/m^3) of Aerosols of Metals and Their Compounds

Substance	Concentration
Aluminum, aluminum oxide, alloys	2.0
Beryllium and its compounds	0.001
Vanadium and its compounds:	
fumes of vanadium pentoxide	0.1
dust	0.5
ferrovanadium	1.0
Tungsten, tungsten carbide	6.0
Germanium, germanium oxide	2.0
Ferric oxide with admixture of fluorine or manganese compounds and free silica (up to 10%)	4.0
Iron oxides (scale)	6.0
Reduced iron	6.0*
Iron with graphite addition (up to 3%)	6.0*
Iron with nickel addition (up to 30%)	1.0*
Carbonyl iron V-3 and R-10	4*
Carbonyl iron R-10 coated with SiO_2	2*
Cadmium oxide	0.1
Cobalt oxide	0.5
Silicon-copper alloy	4.0
Manganese (expressed as Mn_4O_2)	0.3
Copper	0.5*
Cupric oxide	0.1*
Copper with graphite (up to 3%), tin (up to 10%), and nickel (30%) additions	0.5*
Soluble molybdenum compounds as:	
condensation aerosol	2.0
dust	4.0
Insoluble molybdenum compounds	6.0
Arsenic pentoxide and trioxide	0.3
Nickel, nickel oxide	0.5
Lead and its inorganic compounds	0.01
Amorphous selenium	2.0
Selenium dioxide	0.1
Mercuric chloride	0.1
Thallium iodide, bromide	0.01
Tantalum oxides	10.0
Tellurium	0.01
Titanium oxides	10.0
Thorium	0.05
Soluble uranium compounds	0.015
Insoluble uranium compounds	0.075
Chromium trioxide, chromates, dichromates (expressed as Cr_2O_3)	0.1
Chromium	2.0
Ferrochromium	2.0
Zinc oxide	5.0
Metallic zirconium and its insoluble compounds (zircon, dioxide, carbide)	5.0
Fluorozirconate	1.0

*Recommended on the basis of our investigations.

Maximum Allowable Concentrations (mg/liter) of Some Volatile and Gaseous Substances

Substance	Concentration
Benzine	0.3
Benzene	0.02
Carbon monoxide	0.02
Nitrogen oxides	0.005
Sulfuric acid, sulfur trioxide	0.001
Sulfur dioxide	0.01
Ethanol	1.0
Hydrocarbons (expressed as C)	0.3

Maximum Allowable Concentrations (mg/m^3) of Dust of Refractory Compounds, Chalcogenides, and Carbonyls

Compound	Formula	MAC	Compound	Formula	MAC
Boride of			Nitride of		
titanium	TiB_2	2.0	silicon	Si_3N_4	6.0*
zirconium	ZrB_2	2.0	titanium	TiN	4.0*
chromium	Cr_3B_2	1.0	zirconium	ZrN	4.0*
			aluminum	AlN	6.0*
Boron carbide	B_4C	2.0	niobium	NbN	10.0*
Carbide of			Sulfide of		
silicon	SiC	5.0*	molybdenum	MoS_2	6.0
titanium	TiC	10.0*	tungsten	WS_2	6.0
zirconium	ZrC	5.0	Selenide of		
chromium	Cr_3C_2	4.0	niobium	$NbSe_2$	4.0
tungsten	WC	6.0*	molybdenum	$MoSe_2$	4.0
			tungsten	WSe_2	2.0
Silicide of			Telluride of		
titanium	$TiSi_2$	4.0*	molybdenum	$MoTe_2$	0.1
molybdenum	$MoSi_2$	4.0*	tungsten	WTe_2	0.01
tungsten	WSi_2	6.0*	Carbonyl of		
Boron nitride	BN	6.0*	molybdenum	$Mo(CO)_6$	1.0*
Boron carbonitride	BNC	10.0*	tungsten	$W(CO)_6$	2.0*

*Approved by the USSR State Sanitary Inspection Office.

General Requirements Imposed on Industrial Organization

For a radical improvement of the working conditions at powder metallurgy plants it is desirable that standard designs of production buildings and equipment should be developed with allowance for modern hygienic requirements, mechanization and automation of processes, creation of production lines, etc. To ensure the most rational use of equipment it is recommended to specialize plants with respect to types of products being manufactured.

Powder metallurgy plants should be accommodated in specially equipped buildings with natural and artificial lighting. The layout of the buildings should provide for maximum flow of the production process. It is recommended to locate the milling-and-preparing department in multistory buildings with the equipment arranged in the production line according to a vertical scheme. The heat-treatment department should be located in one-story buildings or in the upper stories of multistory buildings equipped with ventilating clerestories with consideration of minimum insulation. The perimeter of the buildings should be built on not more than 40% of the total length of the outside walls. The height of the shops should be at least 6 m.

The walls and inside structure of the buildings must be made of materials whose surface is easily cleaned of dust. The floors should be made waterproof with devices for hydraulic removal of dust.

The exhaust ducts transporting the dust-laden air should be vertical or with an inclination to the horizontal exceeding the angle of repose of the material. If it is necessary to use air ducts with a smaller angle of inclination, the velocities of the air should be selected so as not to allow settling of the dust – at least 15 m/sec, and in horizontal sections at least 20-25 m/sec. The dust -laden air being removed from the rooms should be purified before being released into the atomosphere. The incoming air should be delivered to the upper zone of the room with a velocity greater than 0.3 m/sec to preclude its motion in the working zone.

The heat-treatment departments must be equipped with local ventilation to remove gaseous combustion products of gases ignited on spark plugs and the dust and gases from the furnace ports. In addition, general ventilation to remove heat should be provided. This is accomplished through ventilating clerestories or ventilation shafts with a natural or mechanical exhaust. Air conditioning should be provided in the working zone near the charging ports of the electric furnaces.

Equipment which is a source of vibration (motors, engines, crushers, mills, presses) must be installed on separate foundations not connected with the foundations of the buildings and other structures. The housing of the milling-mixing equipment (crushers, mills, mixers, etc.) should be covered by noise-absorbing materials or be installed in noiseproof chambers. The noise level in the working zones should not exceed 85 dB, and when operating presses, 75 dB.

The industrial rooms at powder metallurgy plants must be cleaned by the wet or pneumatic method. For pneumatic removal of dust it is necessary to use a suction system or other equipment providing regular cleaning of the rooms.

Workers exposed to dust (those involved in reduction, milling, carbide production, mixing, proportioning, pressing, sizing, mechanical repairs, classifying, inspecting, etc.) must be provided with head-gear, covered with coveralls, and with ShB-1 "Lepstok" dust-tight respirators or others. The workers of the heat-treatment shops, and also those working near the gas-conversion, powder reduction, and sintering furnaces who are exposed to radiant and convective heat, should be provided with protective goggles, protective shoes, and caps, and covered with protective clothing and gauntlets. If the noise exceeds the allowable levels, the workers must be supplied with ear protectors.

Production of Sintered Articles from Refractory Compounds

The production of refractory compounds is similar in many ways with respect to the equipment used and manufacturing operations to the initial stages of the manufacture of sintered articles. In view of this, the hygienic requirements imposed on the production processes and equipment for the shops can be the same and amount to the following: Milling, screening, transporting, and reloading the powder materials and other processes accompanied by the liberation of dust should be accomplished using sealed equipment and transport means and with local exhaust ventilation.

Ball and vibrating mills and other milling and mixing equipment, their charging devices, and the spouts for unloading the materials, and also parts of conveyer lines adjacent to them, should be covered and provided with exhaust ventilation. It is recommended to weigh the materials on automatic proportioners. The vibrating screen and other equipment for classifying materials should be interlocked with the crushing and milling equipment. The hoppers for the prepared mixtures should be connected hermetically with the discharge openings of the milling-and-mixing equipment and vibrating screens.

It is recommended to mix powdered materials with plasticizers (benzine, ethanol, etc.) in hermetically closed mixers equipped with automatic proportioners. The prepared mixtures should be stored in closed containers and in special lockers equipped with exhaust ventilation.

Articles should be pressed on automatic and semiautomatic presses equipped with local exhaust ventilation and devices preventing hand injuries. It is necessary to provide a continuous supply of the charge during the work shift. When the charge is proportioned and packed manually, all operations should be done under an exhaust hood. Inspection and cleaning of the intermediate products should be done at special stations equipped with exhaust devices.

Multichannel and compartment-type furnaces are not recommended for plants under construction and those being planned. In existing shops multichannel furnaces can be used only in the case of using boats made of materials not requiring dressing. It is recommended to sinter articles in continuous furnaces equipped with mechanical charging devices and automatic temperature control. The operation of compartment-type furnaces is allowed if they have a small capacity. To reduce the liberation of heat from furnaces it is necessary to insulate the heated surfaces or provide heat-abstracting screens. The temperature of the outer surface of the insulation and screen, according to the existing sanitary standards, should not exceed 45°C.

So that the heat from the materials and articles after their heat treatment in the compartment-type and other furnaces (not equipped with coolers) does not enter the room, they should be isolated from the working zones and located in areas equipped with exhaust devices to remove the heat. To protect the workers against infrared radiation it is recommended to equip the electric furnaces with lock chambers with side openings for loading the articles. If this is impossible, the open ports of the furnaces should be equipped with protective devices (shields, gates, etc.).

It is desirable to load the materials and articles into boats on special tables equipped with covers and effective exhaust ventilation. It is recommended to unload the finished articles from the boats under specially equipped exhaust hoods with automatic dumping of the boats onto the vibrating screens. The articles should be separated from the packing by means of vibrating screens. Impregnation of the articles with oil or sulfur, their hardening, and other operations must be done in special shelters with effective exhaust ventilation.

In the machining of sintered articles on various machines it is recommended to replace dry machining by wet. The machines must be equipped with local exhaust fans and with shields preventing injury to the workers by metal chips. The allowable exhaust under the covers of the equipment and transport devices is not less than 0.15 kg/m^3, which is attained by removing them from the air passing through the holes or open apertures and slots at velocities not less than 1 m/sec.

The benches for manual cleaning of the articles must be equipped with ventilation hoods; the air velocity should not be less than 1 m/sec. Local air scoops must be installed near the presses which provide an air velocity of at least 0.6 m/sec on the press platen from the working side. The recommended air velocity in the suction hole of the housing of the abrasive wheels of the cleaning machines is not less than 30% of the peripheral velocity of the wheel.

Production of Iron Powders by Reduction

In the manufacture of iron powders by the reduction method, it is desirable, from a hygienic standpoint, that the scale should be reduced in horizontal periodic-action continuous furnaces [50].

Scale, carbon black, and other granular materials should be transported by the closed method (closed cars, kraft-pulp bags, tank cars). They should be unloaded mechanically.

In pitch storage and in operations accompanied by the liberation of dust and its gaseous products, it is necessary to observe the pertinent requirements and sanitary rules approved by the USSR SSIO. The charge materials should be stored in closed warehouses which should be equipped with lifting and transporting means for the mechanized handling, unloading, and transporting of materials. The exits of the warehouses should be equipped with gates interlocked with air curtains operating when the gates are open.

The delivery of scale and other raw materials to the receiving hoppers of the drying furnace and their transport to the milling and mixing equipment in the charge-preparation department should be mechanized with maximum sealing of the transporting devices.

Processes related with preparation of the charge (milling, mixing, classifying) and the production of iron powder must be incorporated into the production, which calls for mechanized transport of the material, automatic feeding of the equipment, automatic unloading, and packing of the iron powder with attention to the sealing of the equipment and exhaust ventilation of the dust-producing areas.

The hopper for the prepared charge should be equipped with automatic volume proportioners. Unloading of the carbon black should be done by means of automatic unloading machines equipped with covers and exhaust devices at places where the container is returned and the black delivered. Transport of black to the charge hoppers should be done by means of elevators or other devices under a negative pressure. The furnaces for drying the raw materials (scale, etc.) must be insulated and equipped with automatic devices for measuring and regulating the temperature. Emptying, cleaning, lubricating, and loading the trays with the charge should be accomplished automatically. Charging the trays into the furnace and transporting and unloading the iron sponge must be mechanized.

The reduction of scale should be done in horizontal continuous furnaces having instruments for automatic temperature control. The furnaces should be insulated, and the surface temperature of the insulation should not exceed 45°C. It is recommended to cover the charging and unloading ports with remote-controlled gates. The inspection windows must be equipped with protective transparent shields.

It is desirable to use exhaust ventilation to reduce the air temperature and dust content in the working zones of the furnace operators (during charging and unloading, during observations of the reduction regime, when measuring the furnace temperature). It is necessary to install ventilating clerestories over the furnaces to remove heat and gases.

Two zones of incoming air ducts should be used to improve aeration of the heat-treatment departments. The incoming air should be supplied by mechanical ventilation systems in areas between the furnaces. The air being delivered should be conditioned in accordance with the requirements of SN-245-63.

It is recommended the transport of iron sponge to the crushing and milling equipment be mechanized, with consideration given to sealing of the transport means; its pulverization should be done in crushers, crusher rolls, and other devices equipped with well-ventilated and soundproof covers. Ball and vibrating mills and others should be airtight and equipped with soundproofing, automatic feeders, and unloading devices. The screening areas should be equipped with hermetic covers.

When packing iron powder into containers they must be attached hermetically to the hopper or other equipment by means of flexible sleeves. Places of loading the powder into containers should be under exhaust ventilation.

When iron scale is reduced in shaft-type furnaces the processes of loading and unloading the briquettes from the furnaces and from the containers with the charge should be mechanized. The working posts of the personnel manning the control panel of the reduction processes should be housed in a special insulated booth equipped with air conditioning. The covers at places of access to the components of mechanisms for inspecting, lubricating, regulating, and adjusting should be easily removable. If the operation of equipment must be constantly checked, the covers on the section being observed must be made of transparent materials.

In all industrial rooms it is necessary to install plenum-exhaust ventilation which should be interlocked with the devices of the operating equipment. In the production of converted gas the delivery of natural gas to the scrubber, the vapor–gas mixture to the conversion furnaces, and the converted gas to places of consumption should be accomplished through sealed lines. The conversion furnace must be airtight, and the temperature of the outside surface of its jacket should not exceed 45°C. The temperature must be regulated automatically. It is recommended to overhaul the conversion furnace after complete cooling.

Production of Carbonyl Iron Powders

In the production of carbonyl iron powders it is recommended to locate the units for decomposition of iron pentacarbonyl in open, well-ventilated areas [50]. The workers operating these units should be supplied with special, heated rest rooms located between the unit and the main room of the shop. Tanks with the pentacarbonyl and cylinders with carbon dioxide, ammonia, and other gases should be located in a shelter equipped with effective exhaust ventilation and isolated from the shop room. The places where samples are taken for analyzing the gas during decomposition of iron pentacarbonyl and for other purposes should be equipped with automatic gas analyzers hermetically fitted to the general circuit of the unit.

The carbonyl iron should be unloaded from the hatches into receivers hermetically connected with the hoppers during discharge; the latter should be under a slight vacuum to remove the gases and prevent sucking in of the iron powder. It is recommended to unload the materials from the hopper periodically by means of automatic devices.

The whole unit must be cleaned of iron powder and other substances by means of automatic air-suction systems. Transport of iron powder should be mechanized. Carbonyl iron powder must be stored in a specially equipped, ventilated room.

It is recommended that iron powders be classified in a special room. Local exhaust ventilation must be installed at the places of loading and unloading the materials from the classifiers.

Loading of the classifiers, ball mills, and other equipment, and their unloading, should be accomplished by means of special containers tightly connected with the receiving hopper during loading and unloading. Milling, screening, reloading, transporting and other operations accompanied by considerable dust liberation should be done in airtight apparatus. It is recommended that the transport mechanisms be made maximally airtight and be kept under vacuum as far as possible. Loading and unloading of the containers for the reduction of carbonyl powders must be done in separate rooms equipped with local exhaust devices.

The maximum allowable concentrations of the dust of carbonyl iron powders in working zones should correspond to those indicated earlier.

Some Recommendations on Performing the Main Production Processes Accompanied by Dust Liberation

Few constructive decisions have as yet been made to meet hygienic requirements, since the majority of production processes are still in the development stage. The hygienic literature describes various sanitary-engineering devices, verified by practice, that can probably be used also at powder metallurgy plants.

The unsatisfactory state of the air environment in industrial rooms is due primarily to imperfection of the production process, particularly to the absence of mechanization in many production areas and numerous places of reloading the materials, whose transport is usually done manually.

The rational layout of equipment with its incorporation into a single closed line corresponding to an assembly-line production process greatly reduces airborne dust.

The inadequate output of the existing plenum systems with an unregular distribution of the incoming air leads to its undesirable movement within and between rooms with different levels of airborne dust.

A major cause of high dust content of the air is the poor efficiency of the local exhaust devices and their low rates of exhaustion in the working apertures of covers, hoods, and canopies (of the order of 0.1-0.4 m/sec). This also promotes to a considerable extent the liberation of vapors, gases, and heat. In some areas high dust concentrations are detected in the breathing zone of the workers despite high rates of exhaustion (for example, 0.8-0.85 m/sec). Therefore, not only a rational location of equipment but also its mandatory encapsulation is necessary. An example of poor organization of exhaust ventilation is the location of inlet vents over places of dust liberation; the air—dust flow then rises to higher levels, and only after their velocity falls are the dust particles trapped in dust extractors or other devices. Such a direction of dust flow is unnatural and irrational, since the dust particles try to occupy the lowest levels.

As a result of the air-dust flow being directed into the upper zone, an obstacle to settling of the dust particles is created. Therefore, the suction apertures of the local exhaust devices must be located at the level of dust liberation with the direction of the air-dust flow to lower zones, which will permit more effective trapping of large as well as small dust particles, bypassing the breathing zone of the workers. It is also necessary to take into account the correspondence between the diameter and rational form of the suction vent and the configuration of the dust cloud. At many plants of small output the materials are loaded and unloaded from the milling and mixing equipments manually, which at present is the main cause of high air dust content at powder metallurgy plants.

The solution of the problem of combatting dust liberation under the conditions described will probably allow using the vacuum—air system which is being used successfully in the production of arsenic salts, sodium fluorosilicate, and antimony pentasulfide. With vacuum unloading of the milled product from small ball mills [197] it is drawn off by suction by means of fans or vacuum pumps. Such a vacuum–air system can be used for collecting materials from containers and removing them from a mill. RMK water-ring vacuum pumps are installed in the vacuum system as boosters, but the use of devices of this type for hauling metallic materials requires special development in relation to the specific weight and dispersed state of the materials being aspirated.

A considerable source of dust formation in the working zone is the classification of powdered materials. This operation is usually performed with the use of various screens which are rather difficult to cover completely. The use of the air-separation method completely eliminates the liberation of dust into the industrial room.

A sufficiently reliable dust-free method of unloading railroad cars has yet to be found, but some methods of improving it are described in the literature. At plants producing reinforced-concrete articles the closed cars are unloaded by means of vacuum-air devices. A receiving unit with a motor actuating special "claws" which scrape the cement toward receiving holes has been designed and manufactured. It was found, however, that it is more rational to use receiving devices with double tubular nozzles, such as are employed for unloading grain from ship holds.

When unloading hoppers the materials quite frequently become suspended in the narrow hole. To avoid this, compressed air in various mechanical devices – agitators and vibrators – are presently used widely. In the case of using compressed air the lower part of the hopper is lined with porous ceramic tiles under which the compressed air is delivered.

A most important condition for dust-free unloading of granular materials is the maximum reduction of their drop height. In connection with this, the spout is lowered to the bottom of the container and raised as the container fills. It is not recommended to use compressed air to increase the rate of unloading, and instead it is better to employ various screw- or rotating-vane mechanical agitators which would create a uniform release of the material.

The device proposed by Matsak [196] can probably be used for loading containers in powder metallurgy practice. It consists of a tube for delivering the material, which has a twin, concentric

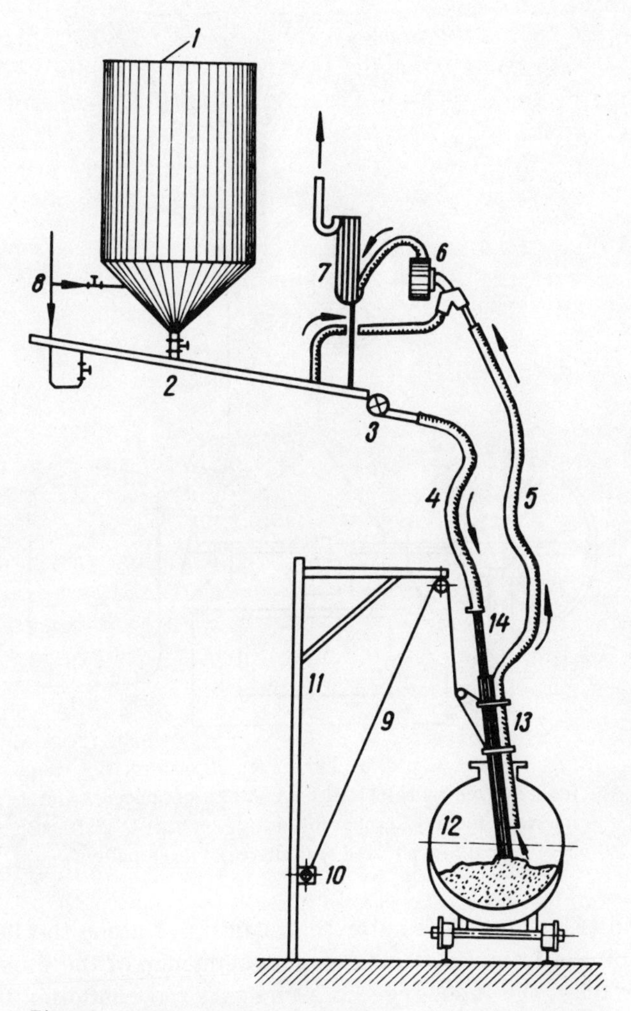

Fig. 40. Diagram of device for loading powdered materials into a container (after Matsak [196]): 1) hopper; 2) pneumatic chute; 3) rotating-vane feeder; 4) tube for delivering materials; 5) suction lines; 6) centrifugal fan; 7) filter bag; 8) delivery of compressed air; 9) cable for raising and lowering tube delivering material; 10) winch; 11) pole with angle brace; 12) receiver; 13) tube (double) for delivering material (air line connected to outer tube); 14) limiter of the movement of the outer tube.

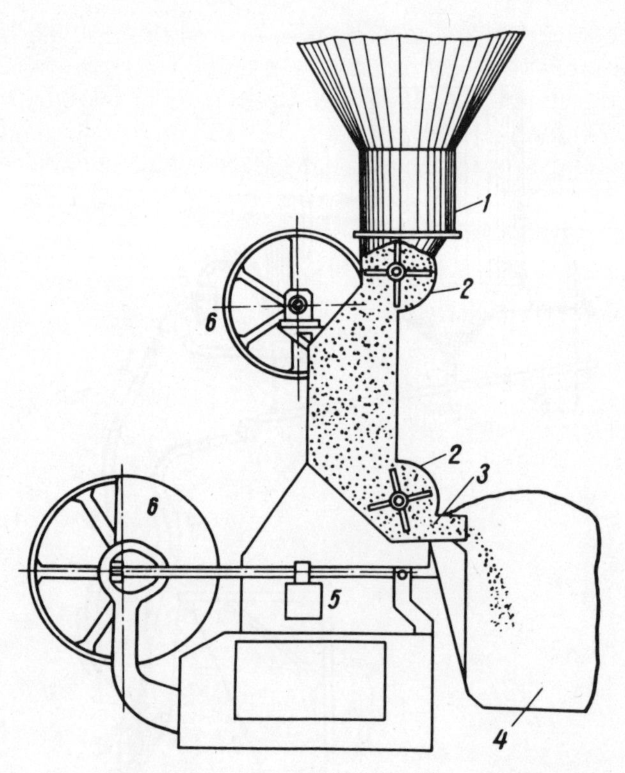

Fig. 41. Diagram of packing machine with rotating-vane feeder (after Matsak [196]): 1) bottom of hopper; 2) rotating-vane feeder; 3) spout for putting on bag; 4) kraft-pulp bag; 5) scales connected with drive; 6) drive pulleys.

lower end (Fig. 40). The outer tube can move along the inner and serves for the inflow of air. As a consequence of the mobility of the outer tube, the discharge hole rises as the container is filled, as a result of which the suction hole of the air line attached to the outer tube, changing its position, is always next to the surface of the dust-producing material. In this case the suction volume is negligible, since only that quantity of air is removed from the container which is displaced from it by the product being loaded.

The area where the metal powders are packed into cans or other containers is distinguished by high dust formation.

To control the formation of dust when packing powdered materials it is best to use special packing machines with a rotating-vane feeder (Fig. 41), which have proved themselves in the cement industry [196, 197]. Other devices have been developed in which the discharge hole is next to the bottom of the container before the beginning of loading, as the container is filled, it descends, and on reaching a prescribed weight, delivery of the product stops (Fig. 42). The liberation of dust can be avoided by means of this device even without local exhaust ventilation [197].

For packing metal powders we can also recommend the packing machine developed by Kasabov for loading toxic substances

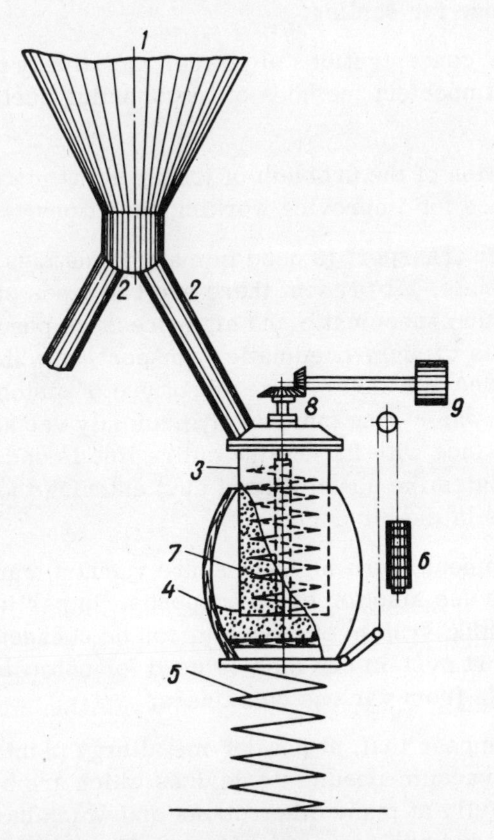

Fig. 42. Diagram of packing machine with discharge holes next to bottom of container (after Matsak [196]): 1) bottom of silo or hopper; 2) spouts; 3) cylinder of packing machine; 4) receiver; 5) rising base; 6) counterweight; 7) vertical screw; 8) bevel gear; 9) drive pulleys.

[196]. A corrugated steel drum with a 120-mm-diameter hole in its top cover is used as the container in this case. After the drum is filled the hole is covered and sealed. The packing machine is equipped with a feeder which ends with a flange pressing tightly against the top cover of the drum. Around the hole of the feeder through which the product enters the drum there are small holes connected to a suction line. The drum is raised and pressed against the flange of the packing machine by means of a prop raised by a lever. When the drum has been filled it is lowered and an asphalt-coated cover placed on the hole. The closed drum is again pressed against the flange for sealing.

High dust concentrations at powder metallurgy plants can be largely due to imperfect methods of transporting dust-producing materials.

The solution of the problem of transporting materials is of great importance for improving working conditions.

Pneumatic transport is used in many industries for moving granular materials. At present there are two types of pneumatic transport: vacuum-pneumatic and high-pressure pneumatic. The most effective is vacuum-pneumatic transport, in which the vacuum reaches 5000 mm H_2O as a result of use of piston vacuum pumps or RMK water-ring pumps. Hygienically vacuum transport is preferable, since with its use the entire line is under a vacuum and thus eliminates the possibility of dust entering the room through unsealed places in the air ducts.

The equipment of the high-pressure vacuum transport system permits its use also for other purposes. In particular, the pneumatic cleaning system of the room can be connected to it. The vacuum transport system can also be used for unloading dust-producing materials from various containers.

It would appear that, at powder metallurgy plants, too, it is best to use the vacuum-pneumatic devices which are being employed successfully at many other plants and which have received a good sanitary and hygienic evaluation.

Medical Service for Workers

Literature data on the effect of various metallic aerosols on workers, studies of morbidity, and results of clinical examinations of workers at powder metallurgy plants indicate that the dust of various metals and alloys can cause pathological changes in workers exposed to its action.

In addition to affection of the air passages and the occurrence of feeble pneumoconitic changes, which should be regarded as benign, the parenchymatous organs can be affected, particularly the functions of the liver and kidneys, cardiovascular system, and digestive organs, and changes in the activity of the nervous system are also possible.

One of the preventive measures against occupational diseases is the early detection of disorders in the state of health of workers by means of periodic medical examinations, which should be performed at least once a year in conformity with Order No. 400 of the Ministry of Health of the USSR, dated May 30, 1969. When conducting medical examinations, according to a procedural instruction [146], the participation of a therapist, neuropathologist, and otolaryngologist should be considered mandatory; in certain cases a dermatologist, ophthalmologist, and other specialists should participate in medical examinations.

In view of the fibrogenic effect of metal powders and the danger of pneumoconiotic changes, the lungs should be x-rayed and the function of external respiration investigated.

To reveal disorders in the activity of the cardiovascular system it is necessary to measure the blood pressure, preferably by means of an oscillograph which provides a high accuracy of results. In this case it is recommended to measure the temporal pressure, since in the opinion of many authors it reflects hypo- or hypertension in the cerebral vessels. It is also recommended to measure the blood pressure on both brachial arteries, since the earliest disorders can be manifested by asymmetry of the arterial pressure.

To evaluate the functional state of the myocardium in the case of appropriate indications, it is necessary to conduct electrocardiography.

The condition of the liver must be determined not only by physical methods of investigation but also by means of functional tests (determination of urobilin in the urine, Quick's test, determination of aminotransferase activity, etc.).

When establishing the neurological status it is necessary to use the orthoclinostatic test in addition to the usual investigations and determine the state of the oculocardiac reflex.

Olfactory dysfunction is revealed not only in the usual examination but also by means of Dubrovskii's tono-olfactometer or other method, and by determining the reflex senstivity of the mucous membrane of the nose and pharynx after tactile stimulation.

In persons working a long time in contact with toxic metallic aerosols it is necessary to investigate carefully the state of the stomach and kidneys and changes in the peripheral blood. Allergic conditions of examinees and presence of dermatitides and other skin lesions should also be taken into account.

When disorders are found in the state of health of persons, it is necessary to conduct outpatient or inpatient treatment with the use (according to the indications) of appropriate antidote and diet therapy, and also physiotherapy methods, or send the patients to sanitariums or rest homes to improve their health.

In the case of frequently aggravated chronic diseases of the air passages, bronchi, and lungs, asthmoid states, organic diseases of the cardiovascular system, diseases of the digestive organs, kidneys, and dermatitides and other serious skin lesions, the workers must be removed from their contact with metal powders. Organic diseases of the central nervous system, neurotic states, and diseases of the optic nerve and retina are also contraindications for continuing to work at powder metallurgy plants.

However, the decision to transfer to another job should be made individually, with consideration of the expressivity of the pathological state, age, service, and specific sanitary and hygienic characteristics of the working conditions at the worker's location.

In a number of cases a brief (1-2 weeks) change to another job is sufficient for recovering health. In this case, to compensate for any decrease in earnings the medical consultation commission draws up a supplementary-payment sick list according to which the difference in wages is made up from the social insurance funds.

According to sanitary standards SN-245-63 [313], a health center of category IV (with one medical assistant) is set up at powder metallurgy plants and in shops with 300 and more workers. If the number reaches 800, a health center of category III (one physician) is created, and for 800-1200 workers – one of category II (two physicians). In the case of a larger contingent of workers a health center staffed by four physicians is organized.

The realization of these recommendations and hygienic requirements on the organization of production processes, arrangement of the rooms and equipment, and timely and thorough medical examinations of the workers will improve working conditions, fortify the health of the workers, and increase their performance.

Conclusion

Powders of metals and their compounds are a source of high dust pollution of the air at industrial plants. The condensation and disintegration aerosols formed during various production operations have a harmful effect on workers.

Persons exposed to metallic dust are affected not only by respiratory diseases but also by dysfunction of the cardiovascular and nervous systems, gastrointestinal tract, liver, and kidneys. Allergic states occur quite frequently.

Results of quantitative experimental evaluations of the fibrogenic activity of various powdered materials permit the conclusion that metals and their compounds have a feeble fibrogenic effect, which in most cases is one-fifth to one-sixth than that of quartz dust.

Only some of the compounds investigated by us (borides, silicides) were distinguished by a slightly greater fibrogenic effect. The fibrogenic activity of powdered substances decreased somewhat with an increase of the toxic effect, but we did not find a clear-cut correlation between fibrogenic activity and the electronic and crystal structure. In some cases, however, we did observe a decrease of fibrogenic activity in a number of substances with a diamond-like structure, which corresponds to a high statistical weight of the sp^3 configurations (for example, in the case of the effect of boron carbonitride on the organism).

More characteristic for many metals is a chronic, and in some instances acute, toxic effect related with metabolic disorders as a result of blocking of sulfhydryl, amino, and carboxyl groups of enzymes and other proteins, and also as a consequence of disturbance of the biosynthesis of nucleic acids.

A comparison of the toxicity of various substances with the character of their electronic and crystal structures indicates the existence of a correlation. The toxicity of simple substances and compounds increases with an increase of the fraction of delocalized electrons and decrease of the statistical weight of stable electronic configurations. It also increases with a decrease of the degree of symmetry of the crystal lattice. These conclusions make it possible to predict the probable toxic effect of new substances even before their detailed toxicological investigation, allowing precautionary measures to be taken.

A thorough investigation of the effects of powders of metal mixtures and various compounds has enabled us to recommend maximum allowable concentrations of dust in the air of industrial rooms. In view of the discovery of a correlation between toxicity and the electronic and crystal structures of substances it is now possible to recommend approximate MAC's for those substances which have still not been subjected to a toxicological investigation.

Literature data and results of our investigations indicate a possible relation between disturbance of the activity of enzymes, and also of other polypeptides, and electronic structure of these substances. Apparently we can draw a parallel between the capacity of metals for polarization of the active groups of protein molecules and their toxicity. Ions of metals of the fourth period have the maximum polarizability and those of the sixth the minimum; accordingly, the metals of the fourth period (V, Cr, Mn) also have maximum biological activity, whereas Hf, Ta, and W are low-activity metals. Further special investigations are of course needed in this direction.

Preliminary observations indicate that destructability of substances is governed by their electronic and crystal structure and determines the possibility of dust formation during dispersion of the materials under production conditions. This quite important problem is of considerable hygienic significance and requires final solution.

The results of industrial and experimental investigations have enabled us to work out hygienic requirements and recommendations for healthier working conditions, which are the basis of the existing Sanitary Rules for powder metallurgy plants and of procedural instructions concerning prevention of diseases in workers engaged in production of refractory compounds, ferrites, and iron powders. The realization of these requirements will improve the working conditions of the workers and increase their performance.

Literature Cited

1. L. A. Abramovich and A. M. Tabachnik, Sovetskii Vestnik Venerologiya i Dermatologii, 7, 13 (1932).
2. N. V. Ageev, Nature of the Chemical Bond in Metal Alloys [in Russian], Izd. AN SSSR, Moscow-Leningrad (1947).
3. N. N. Anichkov, Study of the Reticuloendothelial System [in Russian], Medgiz, Leningrad (1930).
4. M. A. Aripov, Author's Abstract of Candidate's Dissertation, Tashkent-Med. Inst. (1959).
5. L. N. Arkhangel'skaya, N. A. Zhilova, N. V. Mezentseva, O. Ya. Mogilevskaya, and T. A. Roshchina, in: Industrial Toxicology and Clinical Aspects of Occupational Poisoning of a Chemical Etiology [in Russian], Medgiz, Moscow (1962), p. 169.
6. L. N. Arkhangel'skaya, O. Ya. Mogilevskaya, and S. S. Spasskii, in: Chemical Factors of the External Environment and Their Hygienic Significance [in Russian], Meditsina, Moscow (1965), p. 14.
7. L. N. Arkhangel'skaya, in: New Data on the Toxicology of Rare Metals and Their Compounds [in Russian], Meditsina, Moscow (1967), p. 181.
8. L. N. Arkhangel'skaya, T. A. Bystrova, and S. S. Spasskii, in: New Data on the Toxicology of Rare Metals and Their Compounds [in Russian], Meditsina, Moscow (1967), p. 194.
9. L. N. Arkhangel'skaya and S. S. Spasskii, in: New Data on the Toxicology of Rare Metals and Their Compounds [in Russian], Meditsina, Moscow (1967), p. 189.
10. O. G. Arkhipova, Author's Abstract of Doctoral Dissertation, Institute of Industrial Hygiene and Occupational Diseases AMN SSSR, Moscow (1967).
11. S. I. Ashbel', R. G. Khil', and S. P. Shatrova, in: Industrial Toxicology and Clinical Aspects of Occupational Diseases of Chemical Etiology [in Russian], Medgiz, Gor'kii (1962), p. 299.
12. E. D. Bakalinsksaya, Transactions of the Symposium on the Problem of Pneumoconiosis, May 20-25, 1967 [in Russian], Medgiz, Moscow (1959), p. 82.

13. K. Balhausen, Introduction to Ligand Field Theory [Russian translation], Mir (1964).
14. F. Basolo and R. Johnson, Chemistry of Coordination Compounds [Russian translation], Mir, Moscow (1966).
15. M. M. Batsheva and S. V. Miller, in: Transactions of the Ukrainian Institute of Industrial Hygiene and Occupational Diseases [in Russian], Kharkov, 24: 66 (1940).
16. A. Bek, Electronic Structure of Transition Metals and Chemistry of Their Alloys [in Russian], Metallurgiya, Moscow (1966).
17. G. V. Belobragina, Summaries of the Tenth Scientific Session of the Sverdlovsk Institute of Industrial Hygiene and Occupational Diseases [in Russian] (1960), p. 45.
18. N. A. Belozerskii, Metal Carbonyls [in Russian], Metallurgizdat, Moscow (1958).
19. L. N. Belyaeva, in: Problems of Industrial Hygiene, Occupational Diseases, and Industrial Toxicology [in Russian], Izd. Sverdlovskogo Inst. Gigieny Truda i Profzabolevanii, 3(2):19 (1958).
20. L. N. Belyaeva, in: Problems of Industrial Hygiene, Occupational Diseases, and Industrial Toxicology [in Russian], Izd. Sverdlovskogo Inst. Gigieny Truda i Profzabolevanii, 2:175 (1958).
21. L. N. Belyaeva, in: Industrial Toxicology and Clinical Aspects of Occupational Diseases of Chemical Etiology [in Russian], Medgiz, Moscow (1962), p. 245.
22. G. S. Berezyuk, in: Problems of Industrial Hygiene and Occupational Diseases [in Russian], Leningrad (1967), p. 280.
23. G. S. Berezyuk, Transactions of the Scientific Session of the Leningrad Scientific Research Institute of Industrial Hygiene and Occupational Diseases [in Russian], Leningrad (1958), p. 43.
24. I. B. Bersuker and A. V. Ablov, The Chemical Bond in Complex Compounds [in Russian], Izd. Shtinitsa, AN Moldavskoi SSR, Kishinev (1962).
25. F. Ya. Bernshtein and A. I. Glushakov, Uchebnye Zapiski Vitebskogo Veterinarnogo Inst., 7, 243 (1940).
26. E. V. Biron, Zh. Russ. Fiz. Khim. Obshch., 11-17(4):964 (1915).
27. G. A. Born, A. Z. Kach, and D. I. Semenov, in: Theoretical Problems of Mineral Metabolism [in Russian], Nauka, Moscow (1966), p. 38.
28. G. B. Bokii, Crystal Chemistry [in Russian], Nauka, Moscow (1960).
29. I. T. Brakhnova, in: Problems of Industrial Hygiene and Occupational Diseases [in Russian], Zdorov'e, Kiev (1968).
30. I. T. Brakhnova and I. G. Tkacheva, Proceedings of the 12th Scientific Session of the Sverdlovsk Institute of Industrial Hygiene and Occupational Diseases [in Russian], (1962), p. 17.
31. I. T. Brakhnova, Summaries of Reports of the Fourth Republican Conferences on Problems of the Production of Cermet Articles and Their Introduction into Industry [in Russian], Kiev (1963), p. 711.
32. I. T. Brakhnova, Bol'shaya Meditsinskaya Éntsiklopediya, Vol. 36, Sovetskaya Éntsiklopediya, Moscow (1964), p. 711.
33. I. T. Brakhnova and I. G. Tkacheva, Sanitary Rules for Powder Metallurgical Plants (Draft) [in Russian], Izd. AN Ukr. SSR, Kiev (1964).

34. I. T. Brakhnova, Summaries of Reports of the Third Seminar on Crystal Chemistry of Elements and Compounds [in Russian], Izd. Kievskogo Politekhnicheskogo Inst. i IPM AN Ukr. SSR (1968), p. 51.
35. I. T. Brakhnova, A. P. Golovatyukh, and L. M. Naumenko, in: Hygiene. Proceedings of the Seventh Congress of Ukrainian Hygienists [in Russian], Zdorov'e, Kiev (1964).
36. I. T. Brakhnova, First Scientific Seminar of the Department of Powder Metallurgy and Rare Metals of the Kiev Polytechnic Institute and Department of Refractory Compounds of the Institute of Powder Metallurgy, Academy of Sciences of the Ukrainian SSR [in Russian], (1965), p. 3.
37. I. T. Brakhnova, in: Proceedings of the Scientific Session of the Donetsk Institute of Industrial Hygiene and Occupational Diseases [in Russian] (1965), p. 13.
38. I. T. Brakhnova and G. V. Samsonov, Poroshkovaya Metallurgiya, 9, 101 (1966).
39. I. T. Brakhnova, Fifth Interinstitute Seminar on the Physical Properties and Electronic Structure of Transition Metals, Their Refractory Compounds and Alloys (Summaries of Reports) [in Russian], Izd. AN Ukr. SSR, Kiev (1966), p. 103.
40. I. T. Brakhnova, in: Malaya Meditsinskaya Éntsiklopediya, Vol. 5, Sovetskaya Éntsiklopediya, Moscow (1967), p. 906.
41. I. T. Brakhnova, Toxicity of Metal Powders Used in Powder Metallurgy [in Russian], Izd. IPM AN Ukr. SSR, Kiev (1970).
42. I. T. Brakhnova and E. D. Bakalinskaya, Poroshkovaya Metallurgiya, 3, 105 (1967).
43. I. T. Brakhnova, Means of Preventing Occupational Diseases when Working with Refractory Compounds at Powder Metallurgy Plants (Methodological Note) [in Russian], Izd. MZ Ukr. SSR, Kiev (1967).
44. I. T. Brakhnova, G. E. Zhirnova, and S. A. Mosendz, Means of Preventing Occupational Diseases in Ferrite Production at Powder Metallurgy Plants [in Russian], Izd. MZ Ukr. SSR, Kiev (1969).
45. I. T. Brakhnova, Poroshkovaya Metallurgiya, 9, 97 (1967).
46. I. T. Brakhnova, Gigiena Truda i Profzabolevaniya, 1, 26 (1969).
47. I. T. Brakhnova, Gigiena Truda i Profzabolevaniya, 9, 7 (1968).
48. I. T. Brakhnova, Gigiena i Sanitariya, 9, 97 (1968).
49. I. T. Brakhnova and G. V. Samsonov, Gigiena i Sanitariya, 42 (1970).
50. I. T. Brakhnova and G. E. Zhirnova, Means of Preventing Occupational Diseases in the Production of Iron Powder and Articles of Them at Powder Metallurgy Plants (Methodological Note) [in Russian], Izd. MZ Ukr. SSR, Kiev (1968).
51. I. P. Brunshtein and S. S. Selikhodzhaev, Gigiena Truda i Profzabolevaniya, 2, 50 (1967).
52. L. A. Buldakov, D. I. Moskalev, and D. I. Semenov, Meditsinskaya Radiologiya, 6, 42 (1960).
53. G. I. Bunimovich and N. N. Buravlev, in: Problems of Industrial Hygiene, Pathology, and Industrial Toxicology [in Russian], Izd. Sverdlovskogo Inst. Gigieny Truda i Profzabolevanii, Part 1, 3, 175 (1958).

54. A. I. Burshtein, Gigiena Truda, 7, 25 (1925).
55. V. V. Vasil'eva, K. M. Ladkin, V. E. Plyushchev, and Yu. S. Sklyarenko, in: Pharmacology and Chemistry, Proceedings of the 11th All-Union Conference of Pharmacologists Devoted to the 100th Anniversary of the Birth of N. P. Kravkov [in Russian] (1965), p. 53.
56. A. G. Veretinskaya, M. S. Tolgskaya, and I. V. Pavlova, Gigiena Truda i Profzabolevaniya, 9, 41 (1966).
57. V. I. Vernadskii, Outlines of Geochemistry [in Russian], Geolitizdat, Moscow (1934).
58. A. E. Vermel', Occupational Bronchial Asthma [in Russian], Meditsina, Moscow (1966).
59. R. J. Williams, in: Role of Metals in the Action of Enzymes, Transactions of the 5th International Biochemical Congress, Symposium IV [in Russian], Izd. AN SSSR, Moscow (1962), p. 160.
60. A. O. Voinar, Biological Role of Trace Elements in Animals and Man [in Russian], Medgiz, Moscow (1953).
61. R. S. Vorob'eva and S. V. Suvorov, Gigiena Truda i Profzabolevaniya, 10, 45 (1961).
62. R. S. Vorob'eva, in: Toxicology of Rare Metals [in Russian], Medgiz, Moscow (1963), p. 135.
63. E. I. Vorontsova and T. S. Karacharov, Summaries of Reports at the 13th All-Union Congress of Hygienists, Epidemiologists, Microbiologists, and Infectious Disease Specialists [in Russian], Medgiz, Moscow (1956), p. 244.
64. E. I. Vorontsova, Gigiena i Sanitariya, 4, 24 (1949).
65. E. I. Vorontsova, Gigiena Truda i Profzabolevaniya, 10, 3 (1960).
66. E. I. Vorontsova, Industrial Hygiene of Arc-Welders [in Russian], Medgiz, Moscow (1960).
67. V. A. Gavrilova and S. V. Miller, Industrial Hygiene in Aluminum Metallurgy. Handbook on Industrial Hygiene [in Russian], Vol. 3 (1961), p. 192.
68. I. D. Gadaskina, in: Problems of Industrial Hygiene and Occupational Diseases [in Russian], Izd. Leningradskogo Inst. Gigieny Truda i Profzabolevanii (1967), p. 158.
69. I. G. Gel'man, in: Improvement of Working Conditions and Revolution of the Way of Life [in Russian], No. 6 (1925), p. 13.
70. I. A. Gel'fon, in: Industrial Toxicology and Clinical Aspects of Occupational Diseases of Chemical Etiology [in Russian], Medgiz, Moscow (1962), p. 237.
71. L. A. Glushkov, Elimination of the Dust of the Equipment of Crushing and Milling Departments [in Russian], Izd. Sverdlovskogo Inst. Gigieny Truda i Profzabolevanii (1957).
72. V. M. Goldschmidt, Usp. Fiz. Nauk, 9, 6, 811 (1929).
73. B. I. Gol'dshtein, Effect of Sulfhydryl Groups on the Biological Properties of Tissue Proteins [in Russian], Zdorov'e, Kiev (1955).
74. B. I. Gol'dshtein, in: Thiol Compounds in Medicine [in Russian], Zdorov'e, Kiev (1959), p. 49.
75. B. I. Gol'dshtein and V. V. Gerasimova, Ukr. Biokhim. Zh., 35, 3 (1936).

76. Yu. M. Golutvin, Heat of Formation and Types of Chemical Bond in Inorganic Crystals [in Russian], Izd. AN SSSR, Moscow (1962).
77. V. A. Gorbatyuk, First Scientific Seminar of the Department of Powder Metallurgy and Rare Metals of the Kiev Polytechnic Institute and Department of Refractory Compounds of the Institute of Powder Metallurgy, Academy of Sciences of the Ukrainian SSR (Summaries of Reports) [in Russian], Kiev (1965), p. 10.
78. N. A. Goryunova, Chemistry of Diamond-like Semiconductors [in Russian], Leningrad (1963).
79. L. A. Goryacheva, in: Industrial Toxicology and Clinical Aspects of Occupational Diseases of Chemical Etiology [in Russian], Medgiz, Moscow (1962), p. 284.
80. L. N. Gratsianskaya, I. N. Velikson, and L. É. Gorn, in: Manganese Oxides [in Russian], Medgiz. Leningrad (1962).
81. V. K. Grigorovich, Mendeleev's Periodic Law and Electronic Structure of Metals [in Russian], Nauka, Moscow (1966).
82. V. K. Grigorovich, in: High-Temperature Inorganic Compounds [in Russian], Naukova Dumka, Kiev (1965), p. 5.
83. V. K. Grigorovich, in: Structure and Properties of Heat-Resistant Metallic Materials [in Russian], Nauka, Moscow (1967), p. 57.
84. Z. É. Grigor'ev, in: Problems of Industrial Hygiene and Occupational Diseases [in Russian], Izd. Leningradskogo Inst. Gigieny Truda i Profzabolevanii (1967), p. 45.
85. A. V. Grinberg, I. D. Makulova, and N. A. Semenovskaya, in: Problems of Industrial Hygiene and Occupational Diseases [in Russian], Izd. Leningradskogo Inst. Gigieny Truda i Zabolevanii (1967), p. 36.
86. A. A. Grinberg, Introduction to the Chemistry of Complex Compounds [in Russian], Goskhimizdat, Moscow (1967).
87. A. V. Grinberg and O. G. Zolotokrylina, in: Transactions of the Jubilee Session Devoted to the 30th Anniversary of the Leningrad Institute of Industrial Hygiene and Occupational Diseases (1924-1954) [in Russian], Leningrad (1957).
88. Ya. M. Grushko, Chromium Compounds and Prevention of Poisoning by Them [in Russian], Meditsina, Moscow (1964).
89. A. G. Gul'ko, Gigiena i Sanitariya, 11, 24 (1965).
90. P. P. Dvizhkov, Transactions of the Saratov State Medical Institute [in Russian], No. 29 (1960), p. 129.
91. P. P. Dvizhkov, Pneumoconioses (Etiology, Pathological Anatomy, Pathogenesis) [in Russian], Meditsina, Moscow (1965).
92. D. M. Dzhangozina and L. I. Slutskii, Gigiena Truda i Profzabolevaniya, 1, 18 (1968).
93. S. R. Dikhtyar, in: Problems of Industrial Hygiene and Occupational Diseases in Nonferrous Metallurgy [in Russian], Izd. Severo-Osetinskogo Med. Inst., Ordzhonikidze (1967).
94. S. R. Dikhtyar, Z. I. Izraél'son, V. A. Litkens, and S. V. Miller, Gigiena Truda i Prozabolevaniya, 11, 49 (1967).
95. E. V. Donat, Elimination of Dust in the Production of Metal Powders [in Russian], Proftekhizdat, Moscow (1958).

96. I. M. Dyatlova, Author's Abstract of Doctoral Dissertation, Institute Reaktivov, Moscow (1967).
97. Yu. A. Egorov, in: Toxicology of Rare Metals [in Russian], Medgiz, Moscow (1963), p. 95.
98. Yu. A. Egorov, in: Toxicology of Rare Metals [in Russian], Medgiz, Moscow (1963), P. 110.
99. V. G. Eliseev, Connective Tissue, Medgiz, Moscow (1961).
100. S. L. Eolyan, O. A. Akopyan, M. A. Avanyan, and S. A. Ovsepyan, in: Industrial Toxicology and Clinical Aspects of Occupational Diseases of Chemical Etiology [in Russian], Medgiz, Moscow (1962), p. 244.
101. V. P. Ershov, Author's Abstract of Candidate's Dissertation, Inst. Gigieny Truda i Profzabolevanii AMN SSSR, Moscow (1962).
102. V. P. Ershov and S. S. Lipinskii, in: New Data on the Toxicology of Rare Metals and Their Compounds [in Russian], Vysshaya Shkola, Moscow (1966), p. 63.
103. R. K. Zhakenova, Author's Abstract of Candidate's Dissertation, Inst. Gigieny Truda i Profzabolevanii AMN SSSR, Moscow (1967).
104. N. A. Zheligovskaya and I. I. Chernyaev, Chemistry of Complex Compounds [in Russian], Vysshaya Shkola, Moscow (1966).
105. N. A. Zhilova, in: New Data on the Toxicology of Rare Metals and Their Compounds [in Russian], Meditsina, Moscow (1967), p. 51.
106. G. E. Zhirnova and E. D. Bakalinskaya, Poroshkovaya Metallurgiya, 11, 106 (1967).
107. V. A. Zamanskii and S. L. Sher, Transactions of the Sverdlovsk Institute of Experimental Medicine [in Russian], No. 4 (1940), p. 167.
108. I. P. Zamotaev and Z. M. Mel'nikova, Zdravookhranenie RSFSR, 11(10):37 (1967).
109. B. E. Izraél', in: Problems of Hygiene and Occupational Diseases [in Russian], Izd. AMN SSSR, Moscow (1948), p. 42.
110. L. A. Zorina, in: Industrial Toxicology and Clinical Aspects of Occupational Diseases of Chemical Etiology [in Russian], Medgiz, Moscow (1962), p. 297.
111. Z. I. Izraél'son, Toxicology of Rare Metals [in Russian], Medgiz, Moscow (1963).
112. Z. I. Izraél'son, in: New Data on the Toxicology of Rare Metals and Their Compounds [in Russian], Meditsina, Moscow (1967).
113. Z. I. Izraél'son and S. V. Suvorov, in: New Data on the Toxicology of Rare Metals and Their Compounds [in Russian], Meditsina, Moscow (1967), p. 249.
114. Z. I. Izraél'son and O. Ya. Mogilevskaya, in: Handbook on Industrial Hygiene [in Russian], Vol. 3, Medizdat, Moscow (1961), p. 311.
115. D. Yost, G. Russel, and K. Garner, Rare Earth Elements and Their Compounds [Russian translation], IL, Moscow (1949).
116. M. A. Kazakevich, Klinicheskaya Meditsina, 11, 56 (1948).
117. M. A. Kazakevich, Gigieny Truda i Profzabolevanii, 6, 54 (1960).
118. A. Z. Kalashnikov-Galaiko, Tellurites in the Clinical and Bacteriological Diagnosis of Diphtheria [in Russian], Medgiz, Moscow (1959).

119. Z. S. Kaplun, Author's Abstract of Candidate's Dissertation, I Moskovskii Med. Inst. (1953).
120. Z. S. Kaplun, Farmakologiya i Toksikologiya, 18, 17 (1955).
121. Z. S. Kaplun, Gigiena i Sanitariya, 5, 26 (1956).
122. Z. S. Kaplun and N. V. Mezentseva, in: Toxicology of Rare Metals [in Russian], Medgiz, Moscow (1963), p. 227.
123. Z. S. Kaplun, in: Toxicology of Rare Metals [in Russian], Medgiz, Moscow (1963), p. 164.
124. Z. S. Kaplun and N. V. Mezentseva, in: Industrial Toxicology [in Russian], Medgiz, Moscow (1960), p. 35.
125. B. D. Karpov, in: Problems of Industrial Hygiene and Occupational Diseases [in Russian], Izd. Leningradskogo Inst. Gigieny Truda i Profzabolevanii (1967), p. 175.
126. A. A. Kasparov and N. A. Zhilova, in: New Data on the Toxicology of Rare Metals [in Russian], Meditsina, Moscow (1967), p. 152.
127. A. A. Kasparov, in: New Data on the Toxicology of Rare Metals and Their Compounds [in Russian], Meditsina, Moscow (1967), p. 126.
128. B. A. Katsnel'son, Gigiena Truda i Profzabolevaniya, 2, 24 (1957).
129. T'ai-Ying Kim, in: Toxicology of Rare Metals [in Russian], Medgiz, Moscow (1963), p. 245.
130. R. Kieffer and P. Schwarzkopf, Hard Metals [Translated from the German], Metallurgizdat, Moscow (1957).
131. G. Ya. Klebanov and Edmin, Gigiena Truda, 5, 53 (1927).
132. I. A. Kovalevich, T. S. Karacharov, and G. A. Kochetkova, in: Dust Control in Industry [in Russian], Meditsina, Moscow (1964), p. 150.
133. M. A. Kovnatskii, Clinical Aspects of Pneumoconioses [in Russian], Gosmedizdat, Leningrad (1963).
134. N. P. Kokorev, D. M. Bobrishchev-Pushkin, and V. D. Krantsfel'd, Gigiena Truda i Profzabolevaniya, 10, 23 (1960).
135. I. I. Kornilov, Physicochemical Principles of Heat Resistance of Alloys [in Russian], Izd. AN SSSR, Moscow (1961).
136. I. I. Kornilov, Intermetallic Compounds and Interaction Between Them [in Russian], Nauka, Moscow (1964).
137. I. I. Kornilov and V. V. Glazova, Interaction of Refractory Metals of Transition Groups with Oxygen [in Russian], Nauka, Moscow (1967).
138. A. I. Kosenko, Proceedings of the 11th Plenum of the Republican Commission for Silicosis Control [in Russian], Zdorov'e, Kiev (1959), p. 145.
139. A. I. Kosenko, Gigiena Truda i Profzabolevaniya, 9, 46 (1967).
140. I. Kostov, Crystallography [Russian translation], Mir, Moscow (1965).
141. T. A. Kochetkova, Summaries of Reports of the Symposium on the Problem of Pneumoconioses [in Russian], Moscow (1957), p. 56.
142. E. P. Krapukhina, Gigiena i Sanitariya, 8, 29 (1956).
143. V. D. Krantsfel'd and Z. V. Smelyanskii, Industrial Hygiene in Lead Metallurgy. Handbook on Industrial Hygiene [in Russian], Vol. 3, Medgiz, Moscow (1961), p. 286.

144. B. A. Krivoglaz, V. G. Boiko, et al., in: Industrial Toxicology and Clinical Aspects of Occupational Diseases of Chemical Etiology [in Russian], Medgiz, Moscow (1962), p. 252.
145. B. A. Krivoglaz, V. G. Boiko, et al., Poroshkovaya Metallurgiya, 5, 109 (1962).
146. B. A. Krivoglaz, A. A. Model', et al., Hospital Observation of Powder Metallurgy Workers (Instructional-Procedural Note) [in Russian], Izd. MZ Ukr. SSR, Kiev (1964).
147. Crystal Structures of Arsenides, Sulfides, Arsenosulfides, and Their Analogues [in Russian], Izd. SO AN SSSR, Novosibirsk (1964).
148. E. E. Kriss and K. B. Yatsimirskii, Neorganicheskaya Khimiya, 10(11): 2436 (1965).
149. T. A. Kochetkova, Gigiena Truda i Profzabolevaniya, 11, 34 (1960).
150. Yu. S. Kryukov, in: Problems of Industrial Hygiene and Occupational Diseases [in Russian], Izd. Leningradskogo Inst. Gigieny Truda i Profzabolevanii (1967), p. 44.
151. Yu. S. Kryukov, in: Problems of Industrial Hygiene and Occupational Diseases [in Russian], Izd. Leningradskogo Inst. Gigieny Truda i Profzabolevanii (1967), p. 42.
152. A. A. Kudryavtsev, Chemistry and Technology of Selenium and Tellurium [in Russian], Vysshaya Shkola, Moscow (1961).
153. G. B. Kuznetsov, Data on the Hygienic Characteristics of Copper Dust, Author's Abstract of Candidate's Dissertation, Sverdlovskii Inst. Gigieny Truda i Profzabolevanii (1966).
154. G. B. Kuznetsov, Gigiena Truda i Profzabolevaniya, 3, 22 (1966).
155. A. K. Kul'makhanov, in: Problems of Industrial Hygiene and Occupational Diseases [in Russian], Izd. Leningradskogo Inst. Gigieny Truda i Profzabolevanii (1967), p. 268.
156. É. F. Kuperman, in: Industrial Toxicology and Clinical Aspects of Occupational Diseases [in Russian], Medgiz, Moscow (1962), p. 172.
157. I. L. Kucherovskii and S. M. Yasnitskii, Gigiena i Bezopasnost' Truda, 1, 92 (1933).
158. N. V. Lazarev, Nonelectrolytes [in Russian], Izd. Voennoi Morskoi Med. Akademii, Leningrad (1944).
159. N. V. Lazarev, in: Problems of General Industrial Toxicology [in Russian], Izd. Leningradskogo Inst. Gigieny Truda i Profzabolevanii (1965), p. 7.
160. N. V. Lazarev, in: Industrial Toxicology and Clinical Aspects of Occupational Diseases of Chemical Etiology [in Russian], Medgiz, Moscow (1962), p. 7.
161. N. V. Lazarev, Harmful Substances in Industry [in Russian], Part II, Goskhimizdat, Moscow (1963).
162. N. V. Lazarev and É. N. Levina, Manganeses Oxides [in Russian], Medgiz, Leningrad (1962).
163. É. N. Levina, Author's Abstract of Doctoral Dissertation, Leningradskii Inst. Gigieny Truda i Profzabolevanii (1957).
164. É. N. Levina, Gigiena Truda i Profzabolevaniya, 3, 29 (1957).

165. É. N. Levina and N. A. Minshena, Gigiena i Sanitariya, 8, 27 (1961).
166. É. N. Levina and A. Loit, Gigiena i Sanitariya, 10, 27 (1961).
167. É. N. Levina, in: Harmful Substances in Industry [in Russian], Part II, Goskhimizdat, Moscow (1963), p. 527.
168. É. N. Levina, in: Problems of General and Particular Industrial Toxicology [in Russian], Izd. Leningradskogo Inst. Gigieny Truda i Profzabolevanii (1965), p. 37.
169. É. N. Levina, in: Problems of Industrial Hygiene and Occupational Diseases [in Russian], Izd. Leningradskogo Inst. Gigieny Truda i Profzabolevanii (1967), p. 160.
170. M. L. Levontin, Transactions of the Sverdlovsk Institute of Experimental Medicine, 4, 127 (1940).
171. M. L. Levontin, Transactions of the Sverdlovsk Institute of Experimental Medicine, 4, 149 (1940).
172. V. V. Leonicheva, Gigiena Truda i Profzabolevaniya, 10, 11 (1965).
173. S. S. Lipinskii, Author's Abstract of Candidate's Dissertation, Inst. Gigieny Truda i Profzabolevanii AMN SSSR, Moscow (1959).
174. S. S. Lipinskii, Gigiena Truda i Profzabolevaniya, 4, 47 (1961).
175. I. G. Lipkovich, Arc Welding from an Occupational-Hygienic Standpoint [in Russian] (1933).
176. V. A. Litkens, Industrial Hygiene in Copper Metallurgy. Handbook of Industrial Hygiene [in Russian], Vol. 3, Medgiz, Moscow (1961), p. 239.
177. I. I. Lifshits, in: Transactions of the Jubilee Scientific Session of the Leningrad Institute of Industrial Hygiene and Occupational Diseases [in Russian], (1957), p. 39.
178. Liu Yu-T'an, Voprosy Meditsinskoi Khimii, 7(6):605 (1961).
179. E. I. Lyublina, A. A. Golubev, and A. O. Loit, in: Industrial Toxicology and Clinical Aspects of Occupational Diseases of Chemical Etiology [in Russian], Medgiz, Moscow (1962), p. 33.
180. E. I. Lyublina and V. A. Filov, in: Problems of General and Particular Industrial Toxicology [in Russian], Izd. Leningradskogo Inst. Gigieny Truda i Profzabolevanii (1965), p. 5.
181. E. I. Lyublina, in: Problems of General and Particular Industrial Toxicology [in Russian], Izd. Leningradskogo Inst. Gigieny Truda i Profzabolevanii (1965), p. 7.
182. E. I. Lyublina, in: Problems of General and Particular Industrial Toxicology [in Russian], Izd. Leningradskogo Inst. Gigieny Truda i Profzabolevanii (1965), p. 26.
183. E. I. Lyublina, Gigiena Truda i Profzabolevaniya, 12, 9 (1967).
184. A. M. Makarchenko, Changes of the Nervous System in Manganese Poisoning [in Russian], Zdorov'e, Kiev (1956).
185. M. K. McQuillen, Phase Transitions in Titanium and Its Alloys [Russian translation], Metallurgiya, Moscow (1967).
186. E. I. Makovskaya and E. D. Bakalinskaya, Vrachebnoe Delo, 12, 1203 (1955).

187. Ya. S. Malakhov and G. V. Samsonov, Poroshkovaya Metallurgiya, 12, 84 (1966).
188. E. M. Malevannaya, Vrachebnoe Delo, 12, 121 (1961).
189. D. P. Malyuga, Dokl. Akad. Nauk SSSR, 31(2):145 (1941).
190. V. I. Marchenko and G. V. Samsonov, in: The Chemical Bond in Semiconductors and Solids [in Russian], Nauka i Teknika, Moscow (1965), p. 216. [English translation in: Chemical Bonds in Semiconductors and Solids, N. N. Sirota, ed., Consultants Bureau, New York (1967).]
191. E. I. Matantseva, Gigiena Truda i Profzabolevaniya, 7, 41 (1960).
192. V. S. Matytskaya, in: Problems of Industrial Hygiene and Occupational Diseases [in Russian], Izd. Leningradskogo Inst. Gigieny Truda i Profzabolevanii (1967), p. 25.
193. V. S. Matytskaya, in: Problems of Industrial Hygiene and Occupational Diseases [in Russian], Izd. Leningradskogo Inst. Gigieny Truda i Profzabolevanii (1967), p. 27.
194. N. Ya. Matyukhin, in: Problems of Industrial Hygiene and Occupational Diseases [in Russian], Izd. Leningradskogo Inst. Gigieny Truda i Profzabolevanii (1967), p. 191.
195. Ya. Z. Matusevich, Collection of Scientific Papers of the Leningrad Institute of Industrial Hygiene and Occupational Diseases During the Years of the Second World War [in Russian] (1945), p. 138.
196. V. G. Matsak and M. V. Yakovenko, Collection of Papers of the Central Sanitary-Hygiene Laboratory of the Moscow Health Department [in Russian], OGIZ, Moscow (1938), p. 85.
197. V. G. Matsak, in: Dust Control in Industry [in Russian], Meditsina, Moscow (1964).
198. G. A. Meerson, Kratkaya Khimicheskaya Éntsiklopediya, Vol. 2, Sovetskaya Éntsiklopediya, Moscow (1963), p. 423.
199. N. V. Mezentseva, Author's Abstract of Candidate's Dissertation, I Moskovskoi Med. Inst. (1957).
200. N. V. Mezentseva, Gigiena i Sanitariya, 4, 46 (1957).
201. N. V. Mezentseva, in: Toxicology of Rare Metals [in Russian], Medgiz, Moscow (1963), p. 47.
202. N. V. Mezentseva, E. A. Mel'nikova, and O. Ya. Mogilevskaya, in: Toxicology of Rare Metals [in Russian], Medgiz, Moscow (1963), p. 58.
203. N. V. Mezentseva, O. Ya. Mogilevskaya, and T. A. Roshcina, Gigiena i Sanitariya, 5, 97 (1964).
204. N. V. Mezentseva, in: New Data on the Toxicology of Rare Metals and Their Compounds [in Russian], Meditsina, Moscow (1967), p. 207.
205. N. V. Mezentseva, in: New Data on the Toxicology of Rare Metals and Their Compounds [in Russian], Meditsina, Moscow (1967), p. 172.
206. E. A. Mel'nikova, Gigiena i Sanitariya, 3, 25 (1957).
207. D. I. Mendeleev, Principles of Chemistry [in Russian], Vols. 1, 2, Goskhimizdat, Moscow-Leningrad (1947).
208. D. I. Mendeleev, The Periodic Law [in Russian], Izd. AN SSSR, Moscow (1958).
209. A. Z. Men'shikov, S. A. Nemnonov, and G. V. Samsonov, in: High-Temperature Inorganic Materials [in Russian], Naukova Dumka, Kiev (1965), p. 88.

210. K. V. Migai, Gigieny Truda i Profzabolevaniya, 8, 7 (1961).
211. S. V. Miller, in: Work and Health of Arc Welders [in Russian], Izd. Ukrainskogo Tsentral'nogo Inst. Gigieny Truda i Profzabolevanii, Kharkov, 24, 6 (1940).
212. S. V. Miller and L. Z. Shapiro, in: Work and Health of Arc Welders [in Russian], Izd. Ukrainskogo Tsentral'nogo Inst. Gigieny Truda i Profzabolevanii, Kharkov, 24:40 (1940).
213. S. V. Miller, in: Work and Health of Arc Welders [in Russian], Izd. Ukrainskogo Tsentral'nogo Inst. Gigieny Truda i Profzabolevanii, Kharkov, 24: 52 (1940).
214. S. V. Miller, in: Work and Health of Arc Welders [in Russian], Izd. Ukrainskogo Tsentral'nogo Inst. Gigieny Truda i Profzabolevanii, Kharkov, 24:92 (1940).
215. S. V. Miller and S. S. Kangellarikh, in: Work and Health of Arc Welders [in Russian], Izd. Ukrainskogo Tsentral'nogo Inst. Gigieny Truda i Profzabolevanii, Kharkov, 24: 311 (1940).
216. A. Ya. Mil'shtein and A. M. Baru, Gigiena Truda i Profzabolevaniya, 9, 35 (1960).
217. M. Ya. Mirskii, Summaries of Reports of the Second Moscow City Scientific-Practical Conference of Industrial Sanitation Physicians [in Russian], Moscow (1957), p. 9.
218. E. R. A. Meriwether, Some Problems of Industrial Hygiene and Occupational Disease [Translated from the English], Medgiz, Moscow (1960).
219. R. B. Mogilevskaya and R. Yu. Govoruk, Transactions of the Ukrainian Institute of Pathology and Industrial Hygiene [in Russian], Vol. 11 (1932), p. 33.
220. O. Ya. Mogilevskaya, Gigiena i Sanitariya, 3, 20 (1956).
221. O. Ya. Mogilevskaya, E. A. Mel'nikova, and N. V. Mezentseva, Tsvetnye Metally, 4, 51 (1957).
222. O. Ya. Mogilevskaya, Gigiena i Sanitariya, 9, 18 (1961).
223. O. Ya. Mogilevskaya, N. V. Mezentseva, and T. A. Roshchina, in: Industrial Toxicology and Clinical Aspects of Occupational Diseases of Chemical Etiology [in Russian], Medgiz, Moscow (1962), p. 171.
224. O. Ya. Mogilevskaya, in: Toxicology of Rare Metals [in Russian], Medgiz, Moscow (1963), p. 26.
225. O. Ya. Mogilevskaya, in: Toxicology of Rare Metals [in Russian], Medgiz, Moscow (1963), p. 71.
226. O. Ya. Mogilevskaya, in: Toxicology of Rare Metals [in Russian], Medgiz, Moscow (1963), p. 151.
227. O. Ya. Mogilevskaya and N. T. Raikhlin, in: Toxicology of Rare Metals [in Russian], Medgiz, Moscow (1963), p. 195.
228. O. Ya. Mogilevskaya, in: Toxicology of Rare Metals [in Russian], Medgiz, Moscow (1963), p. 314.
229. O. Ya. Mogilevskaya, Poroshkovaya Metallurgiya, 4, 115 (1964).
230. O. Ya. Mogilevskaya, Gigiena Truda i Profzabolevaniya, 6, 40 (1965).
231. O. Ya. Mogilevskaya and T. A. Roshchina, in: New Data on the Toxicology of Rare Metals and Their Compounds [in Russian], Meditsina, Moscow (1963), p. 213.
232. O. Ya. Mogilevskaya, I. A. Zibireva, and G. M. Khosid, in: New Data on the Toxicology of Rare Metals and Their Compounds [in Russian], Meditsina, Moscow (1967), p. 142.

233. O. Ya. Mogilevskaya, in: New Data on the Toxicology of Rare Metals and Their Compounds [in Russian], Meditsina, Moscow (1967), p. 175.
234. O. Ya. Mogilevskaya, in: New Data on the Toxicology of Rare Metals and Their Compounds [in Russian], Meditsina, Moscow (1967), p. 160.
235. V. G. Mogilevskii, Electromagnetic Powder Clutches and Brakes [in Russian], Énergiya, Moscow (1964).
236. A. M. Monaenkova and K. V. Glotova, Gigiena Truda i Profzabolevaniya, 6, 41 (1963).
237. A. M. Monakov, Transactions of the Sverdlovsk Institute of Experimental Medicine, No. 4, 156 (1940).
238. E. V. Montsevichyute-Eringene, Patofiziologiya i Éksperimental'naya Terapiya, 4, 71 (1964).
239. A. O. Navakatiyan, Respiratory Functions in Pneumoconioses and Dust-Induced Bronchitis [in Russian], Meditsina, Moscow (1967).
240. V. K. Navrotskii, Gigiena i Sanitariya, 6, 29 (1960).
241. I. M. Naumenko, Author's Abstract of Candidate's Dissertation, Kievskii Med. Inst. im. Bogomoltsa (1965).
242. Nguen Van Hap, in: Industrial Toxicology and Clinical Aspects of Occupational Diseases of Chemical Etiology [in Russian], Medgiz, Moscow (1962), p. 295.
243. B. V. Nekrasov, Course of General Chemistry [in Russian], Goskhimizdat, Moscow (1962).
244. V. Ya. Neretin, Data on the Toxicology of Lithium [in Russian], Medgiz, Moscow (1959).
245. V. A. Neronov, Aluminum Borides [in Russian], Nauka, SO AN SSSR, Novosibirsk (1966).
246. N. A. Nesmeyanov, Kratkaya Khimicheskaya Éntsiklopediya, Vol. 2 (1964), p. 439.
247. Z. V. Novokhatskaya, Vrachebnoe Delo, 2, 155 (1960).
248. V. A. Obolonchik, Selenides of Rare Earth Elements [in Russian], Naukova Dumka, Kiev (1965).
249. V. A. Obolonchik, in: Chalcogenides [in Russian], Naukova Dumka, Kiev (1967), p. 26.
250. A. I. Olefir, Vrachebnoe Delo, 9, 20 (1967).
251. L. Orgel, An Introduction to Transition-Metal Chemistry [Russian translation], Mir, Moscow (1962).
252. É. Yu. Ornitsan and N. V. Uspenskaya, in: Problems of Industrial Hygiene and Occupational Diseases [in Russian], Izd. Leningradskogo Inst. Gigieny Truda i Profzabolevanii (1962), p. 272.
253. P. N. Ostrin, S. A. Artemenko, A. N. Popov, and N. F. Kolesnik, Poroshkovaya Metallurgiya, 12, 3 (1966).
254. G. M. Parkhomenko, Gigiena i Sanitariya, 9, 11 (1948).
255. T. S. Paskhina, Instructions on the Determination of Aspartate and Alanine Aminostransferases in Human Blood Serum [in Russian], Izd. Inst. Biokhimii AN SSSR, Moscow (1965).
256. L. Pauling, The Nature of the Chemical Bond [Russian translation], Goskhimizdat, Moscow (1947).

257. V. V. Podosinovskii, Gigiena Truda i Profzabolevaniya, 6, 45 (1965).
258. V. V. Podosinovskii, Gigiena i Sanitariya, 1, 28 (1965).
259. A. Polikar and A. Kolle, Physiology of Normal and Pathological Connective Tissue [in Russian], Nauka, SO AN SSSR, Novosibirsk (1966).
260. Occupational Diseases (Handbook for Physicians) [in Russian], Medgiz, Moscow (1957).
261. A. Pullman and B. Pullman, Quantum Biochemistry [Russian translation], Mir, Moscow (1965).
262. B. Pullman, Electronic Biochemistry [Russian translation], Nauka, Moscow (1966).
263. L. V. Rabotnikova, in: Problems of General and Particular Industrial Toxicology [in Russian], Izd. Leningradskogo Inst. Gigieny Truda i Profzabolevanii (1965), p. 52.
264. L. V. Rabotnikova, in: Proceedings of the Tenth Scientific-Practical Conference of Young Hygienists and Sanitary Physicians (June 29-July 2, 1965) [in Russian], Izd. Moskovskogo Inst. Gigieny im. F. F. Érismana (1965), p. 111.
265. S. V. Radzikovskaya, in: Chalcogenides [in Russian], Naukova Dumka, Kiev (1967), p. 18.
266. S. V. Radzikovskaya and V. M. Marchenko, Sulfides of Rare Earth Metals and Actinides [in Russian], Naukova Dumka, Kiev (1966).
267. I. D. Radomysel'skii, Poroshkovaya Metallurgiya, 10, 63 (1967).
268. N. O. Razumovskii and O. L. Gorchinskii, Meditsinskaya Radiologiya, 11, 46 (1960).
269. V. S. Raitses and Ya. V. Ganitkevich, Byull. Éksperim. i Biol. Meditsiny, 48(11):81 (1959).
270. A. M. Rashevskaya and S. V. Levina, Trudy AMN SSSR, Moscow, 16, 78 (1952).
271. L. S. Reznikova, Complement and Its Significance in Immunological Reactions [in Russian], Meditsina, Moscow (1967).
272. H. Remi, Course of Inorganic Chemistry [Russian Translation], Vol. 1, IL, Moscow (1963); Vol. 2, Mir, Moscow (1966).
273. C. Reed, Excited States in Chemistry and Biology [Russian translation], IL, Moscow (1960).
274. B. S. Rodkina, Proceedings of the Scientific Session of the Donetsk Research Institute of Industrial Hygiene and Occupational Diseases [in Russian], Donetsk (1965), p. 160.
275. P. A. Rozenberg and A. A. Orlova, Gigiena Truda i Profzabolevaniya, 12, 33 (1967).
276. I. V. Roshchin, Vrachebnoe Delo, 9, 819 (1952).
277. I. V. Roshchin, Gigiena Truda i Profzabolevaniya, 7, 41 (1961).
278. I. V. Roshchin, in: Industrial Toxicology and Clinical Aspects of Occupational Diseases of Chemical Etiology [in Russian], Medgiz, Moscow (1962), p. 173.
279. A. V. Roshchin, in: Toxicology of Rare Metals [in Russian], Medgiz, Moscow (1963), p. 83.
280. A. V. Roshchin, Author's Abstract of Doctoral Dissertation, I Moskovskii Med. Inst. (1964).
281. I. V. Roshchin, in: New Data on the Toxicology of Rare Metals and Their Compounds [in Russian], Meditsina, Moscow (1967), p. 223.
282. T. A. Roshchina, Gigiena i Sanitariya, 8, 25 (1964).

283. G. Yu. Rozina, L. S. Dubeikovskaya, I. N. Tsiryul'nikova, and É. M. Stepanchenko, Transactions of the Scientific Session of the Leningrad Institute of Industrial Hygiene and Occupational Diseases Based on Results of Studies in 1958 [in Russian], (1959), p. 16.
284. G. I. Rumyantsev, Author's Abstract of Candidate's Dissertation, I Moskovskii Med. Inst. (1955).
285. G. I. Rumyantsev, in: Toxicology of Rare Metals [in Russian], Medgiz, Moscow (1963), p. 176.
286. E. M. Savitskii et al., Alloys of Rare Earth Metals [in Russian], Izd. AN SSSR, Moscow (1962).
287. S. É. Sandratskaya, in: Toxicology of Rare Metals [in Russian], Medgiz, Moscow (1963), p. 117.
288. M. I. Salekhov, Gigiena Truda i Profzabolevaniya, 9, 45 (1962).
289. G. V. Samsonov and Ya. S. Umanskaya, Hard Compounds of Refractory Metals [in Russian], Izd. Metallurgizdat, Moscow (1957).
290. G. V. Samsonov, Silicides and Their Use in Technology [in Russian], Izd. AN Ukr. SSR, Kiev (1958).
291. G. V. Samsonov and L. Ya. Markovskii, Boron, Its Compounds and Alloys [in Russian], Naukova Dumka, Kiev (1960).
292. G. V. Samsonov and K. I. Portnoi, Alloys on a Base of Refractory Compounds [in Russian], Oborongiz, Moscow (1961).
293. G. V. Samsonov and Yu. B. Paderno, Borides of Rare Earth Elements [in Russian], Izd. AN Ukr. SSR, Kiev (1961).
294. G. V. Samsonov and B. M. Tsarev, High-Temperature Cermet Materials [in Russian], Izd. AN Ukr. SSR, Kiev (1961).
295. G. V. Samsonov, Refractory Compounds [in Russian], Metallurgizdat, Moscow (1963).
296. G. V. Samsonov and A. P. Épik, Coatings of Refractory Compounds [in Russian], Metallurgiya, Moscow (1964). [English translation in: Coatings of High-Temperature Materials, H. H. Hausner, ed., Plenum Press, New York (1966).]
297. G. V. Samsonov, Silicides, in: Kratkaya Khimicheskaya Éntsiklopediya, Vol. 4, Sovetskaya Éntsiklopediya, Moscow (1964), p. 866.
298. G. V. Samsonov, Sulfides, in: Kratkaya Khimicheskaya Éntsiklopediya, Vol. 4, Sovetskaya Éntsiklopediya, Moscow (1964), p. 866.
299. G. V. Samsonov, Refractory Compounds of Rare Earth Metals with Nonmetals [in Russian], Metallurgiya, Moscow (1964).
300. G. V. Samsonov and V. Kh. Oganesyan, Dokl. Akad Nauk. SSSR, 10, 1317 (1965).
301. G. V. Samsonov and P. S. Kislyi, High-Temperature Nonmetallic Thermocouples and Sheath [in Russian], Naukova Dumka, Kiev (1965). [English translation: Consultants Bureau, New York (1967).]
302. G. V. Samsonov, V. S. Fomenko, and Yu. B. Paderno, in: High-Temperature Inorganic Compounds [in Russian], Naukova Dumka, Kiev (1965), p. 108.
303. G. V. Samsonov, Role of the Formation of Stable Electronic Configurations in the Formation of the Properties of Chemical Elements and Compounds [in Russian], Izd. IMP AN Ukr. SSR, Kiev (1965).
304. G. V. Samsonov and O. I. Shulishova, in: High-Temperature Inorganic Compounds [in Russian], Naukova Dumka, Kiev (1965), p. 116.

305. G. V. Samsonov, in: High-Temperature Inorganic Compounds [in Russian], Naukova Dumka, Kiev (1965), p. 144.
306. G. V. Samsonov, Ukr. Khim. Zh., 31(10):1005 (1965).
307. G. V. Samsonov, Ukr. Khim. Zh., 31(12):1233 (1965).
308. G. V. Samsonov, Poroshkovaya Metallurgiya, 12, 49 (1966).
309. G. V. Samsonov, Selenides, in: Kratkaya Khimicheskaya Éntsiklopediya, Vol. 4, Sovetskaya Éntsiklopediya, Moscow (1965), p. 783.
310. G. V. Samsonov, Tellurides, in: Kratkaya Khimicheskaya Éntsiklopediya, Vol. 5, Sovetskaya Éntsiklopediya, Moscow (1967), p. 57.
311. G. V. Samsonov, in: Chalcogenides, [in Russian], Naukova Dumka, Kiev (1967), p. 3.
312. G. V. Samsonov, Yu. B. Paderno, and B. M. Rud', Izv. Vuzov. Fizika. Izd. Tomskogo Univ., 9, 129 (1967).
313. Sanitary Regulations of the Design of Industrial Plants SN-245-63 [in Russian], Moscow (1963).
314. I. V. Sanotskii, Farmakologiya i Toksikologiya, 2 (1955).
315. S. S. Selikhodzhaev and Kh. Ya. Vengerskaya, Gigiena i Sanitariya, 10, 78 (1961).
316. D. I. Semenov, in: Theoretical Problems of Mineral Metabolism [in Russian], Nauka, Moscow (1966), p. 5.
317. D. I. Semenov and I. P. Tregubenko, in: Theoretical Problems of Mineral Metabolism [in Russian], Nauka, Moscow (1966), p. 64.
318. A. Szent-Györgyi, Introduction to Submolecular Biology [in Russian], Nauka, Moscow (1964).
319. N. F. Selivanova, First Scientific Seminar of the Department of Powder Metallurgy and Rare Metals of the Kiev Polytechnic Institute and Department of Refractory Compounds of the Institute of Powder Metallurgy, Academy of Sciences of the Ukrainian SSR [in Russian], Kiev (1965), p. 20.
320. N. Z. Slinchenko, Arkhiv Patologii, 6, 21 (1962).
321. N. Z. Slinchenko, Arkhiv Patologii, 2, 94 (1964).
322. L. I. Slutskii and I. I. Sheleketina, Vopr. Med. Khimii, 5(6):466 (1959).
323. L. I. Slutskii and M. I. Érman, Proceedings of the Scientific Session of the Donetsk Inst. of Industrial Physiology [in Russian], (1960), p. 57.
324. A. G. Sobolevskii, Materials in Electronics [in Russian], Gosénergoizdat, Moscow (1963).
325. R. E. Sova, in: Hygiene and Toxicology [in Russian], Zdorov'e, Kiev (1967), p. 165.
326. M. S. Sominskii, Semiconductors [in Russian], Nauka, Moscow (1967).
327. S. N. Sorinson, Data on Problems of Industrial Hygiene and Clinical Aspects of Occupational Diseases [in Russian], Vol. 5, Izd. Gor'kovskogo Inst. Gigieny Truda i Profzabolevanii (1965), p. 133.
328. S. N. Sorinson, Gigiena i Sanitariya, 11, 30 (1957).
329. S. N. Sorinson, A. P. Kornilova, and A. M. Artem'eva, Gigiena i Sanitariya, 9, 69 (1958).
330. N. F. Sosova, Gigiena i Sanitariya, 6, 89 (1960).
331. S. S. Spasskii and L. N. Arkhangel'skaya, in: Chemical Factors of the Environment and Their Hygienic Significance [in Russian], Moscow (1965), p. 19.

332. S. V. Speranskii, Proceedings of the Scientific Session Devoted to the Results of Studies in 1961-1962 (February 5-8, 1963) [in Russian], Izd. Leningradskogo Inst. Gigieny Truda i Profzabolevanii (1963), p. 119.
333. S. V. Speranskii, in: Problems of General and Particular Industrial Toxicology [in Russian], Leningrad (1965), p. 56.
334. A. S. Spirin, Biokhimiya, 23:656 (1958).
335. V. G. Syrkin, New Carbonyl Materials [in Russian], Znanie, Moscow (1965).
336. V. G. Syrkin and I. S. Tolmasskii, Poroshkovaya Metallurgiya, 7, 42 (1965).
337. Ya. K. Syrkin and M. E. Dyatkina, The Chemical Bond and Structure of Molecules [in Russian], Goskhimizdat, Moscow-Leningrad (1946).
338. T. P. Singer and E. B. Kerney, Transactions of the Fifth International Congress [in Russian], Izd. AN SSSR, Moscow (1962), p. 158.
339. V. S. Surat, in: Industrial Toxicology, Clinical Aspects, Hygiene, and Prevention [in Russian], Medgiz, Moscow (1934), p. 45.
340. V. S. Surat, A. P. Sapozhnikov, and A. P. Shilov, Kazanskii Med. Zh., 2, 149 (1936).
341. M. M. Tarnopol'skaya, V. M. Makotchenko, S. M. Bagrova, and V. D. Rozenberg, Gigiena i Sanitariya, 8, 19 (1964).
342. Theoretical Problems of Mineral Metabolism [in Russian], Nauka, Moscow (1966).
343. N. N. Tikhomirov, Gigiena, Bezopasnost' i Patologiya Truda, 11, 12 (1930).
344. I. S. Tolmasskii, Author's Abstract of Candidate's Dissertation, Moskovskii Énergeticheskii Inst. (1962).
345. I. P. Tregubenko, in: Theoretical Problems of Mineral Metabolism [in Russian], Nauka, Moscow (1966), p. 52.
346. I. P. Tregubenko and D. I. Semenov, in: Theoretical Problems of Mineral Metabolism [in Russian], Nauka, Moscow (1966), p. 74.
347. D. N. Trifonov, Rare Earth Elements [in Russian], Izd. AN SSSR, Moscow (1960).
348. A. F. Wells, The Structure of Inorganic Substances [Russian translation], IL, Moscow (1948), p. 347.
349. W. Fyfe, Geochemistry of Solids [Russian translation], Mir, Moscow (1967).
350. I. M. Fedorchenko and R. A. Andrievskii, Principles of Powder Metallurgy [in Russian], Izd. AN Ukr. SSR, Kiev (1963).
351. I. M. Fedorchenko, Poroshkovaya Metallurgiya, 10, 51 (1967).
352. A. E. Fersman, Geochemistry [in Russian], Vol. 3, Khimteoretizdat, Leningrad (1937).
353. I. V. Fetisenko and A. I. Medvedkova, Summaries of Reports of the 8th Interdistrict Conference of Physicists, Biochemists, and Morphologists of the Southeastern RSFSR [in Russian], Voronezh (1948), p. 160.
354. Physicochemical Properties of Elements (Handbook) [in Russian], Naukova Dumka, Kiev (1965).
355. V. S. Filatova, Author's Abstract of Candidate's Dissertation, I Moskovskii Med. Inst. (1948).
356. V. S. Filatova, Farmakologiya i Toksikologiya, 14(4):25 (1951).
357. V. S. Filatova, Gigiena i Sanitariya, 5, 47 (1951).
358. V. A. Filov and E. I. Lyublina, Biofizika, 10(4):602 (1965).
359. I. N. Frantsevich, Powder Metallurgy [in Russian], Znanie, Moscow (1958).

360. I. G. Fridlyand, Handbook on Medical Examinations of Workers in Industry and in Occupations with Harmful Substances [in Russian], Medgiz, Moscow (1950).
361. I. G. Fridlyand, Gigiena i Sanitariya, 8:55, 61 (1959).
362. I. N. Frolova, Summaries of Reports of the Scientific Conference Devoted to the Results of Studies of the Gorkii Institute of Industrial Hygiene and Occupational Diseases in 1956 [in Russian], Gorkii (1957), p. 17.
363. A. A. Khavatasi, Summaries of Reports of the Scientific Session of the Institute of Industrial Hygiene and Occupational Diseases [in Russian], Izd. MZ Gruz. SSR, Tbilisi (1955), p. 13.
364. A. A. Khavatasi, Summaries of Reports of the First Congress of Hygienists and Sanitary Physicians [in Russian], Tbilisi (1956), p. 80.
365. A. A. Khavatasi, Gigiena Truda i Profzabolevaniya, 4, 36 (1958).
366. M. I. Khlebnikova, in: Toxicology of Rare Metals [in Russian], Medgiz, Moscow (1963), p. 278.
367. J. Holum, Molecular Principles of Life [Russian translation], Mir, Moscow (1965).
368. M. P. Chekunova, in: Harmful Chemical Substances in Industry [in Russian], Part II, Vol. 5, Goskhimizdat, Moscow (1963), p. 2.
369. M. P. Chekunova, in: Problems of General and Particular Industrial Toxicology [in Russian], Leningrad (1965), p. 63.
370. M. P. Chekunova, in: Problems of Industrial Hygiene and Occupational Diseases [in Russian], Izd. Leningradskogo Inst. Gigieny Truda i Profzabolevanii (1967), p. 162.
371. D. M. Chizhikov and V. P. Schastlivyi, Selenium and Selenides [in Russian], Nauka, Moscow (1964).
372. D. M. Chizhikov and V. P. Schastlivyi, Tellurium and Tellurides [in Russian], Nauka, Moscow (1966).
373. I. V. Shagan, Investigations in Industrial Hygiene and Occupational Diseases [in Russian], Vol. 75 (1963), p. 181.
374. I. N. Sharkevich, Problemy Éndokrinologii i Gormonoterapii, 5(4):18 (1959).
375. A. M. Shevchenko, Summaries of the Methodological Scientific Session Devoted to the Results of the Studies of the Institute of Industrial Hygiene and Occupational Diseases in 1961-1962 [in Russian], Kiev (1963).
376. É. Shtarkenshtein, É. Pol', and I. Rost, Toksikologiya, 1, 2 (1933).
377. G. S. Érenburg, Poisoning by Copper Compounds and Its Control [in Russian], Izd. Leningradskogo Inst. Gigieny Truda i Profzabolevanii (1938).
378. W. Hume-Rothery, Atomic Theory for Metallurgists [Russian translation], Metallurgizdat, Moscow (1955).
379. Yasnitskii et al., in: Harmful Substances in Industry [in Russian], Part II, Goskhimizdat, Moscow (1963), p. 83.
380. M. L. Amdur, Occup. Med., 3, 386 (1947).
381. A. J. Amor, J. Industr. Hyg., 216-221 (1932).
382. Arrigoni, Med. Lavoro, 8, 12 (1933).
383. J. C. Aub and R. S. Grier, J. Industr. Hyg. Toxicol., 1949, 31, 123.
384. E. Baader, Gewerberkrankungen, Verl. Schwarzenberg, Muchen (1954).
385. A. M. Bactjer, C. Damron, and V. Budaez, Arch. Industr. Health, 14(2):178-188 (1956).

386. O. T. Baily, F. D. Ingraham, P. S. Weadon, and A. F. Susen, J. Neurosurg., 9(1):83 (1952).
387. R. Baldock and J. R. Sitos, Phys. Rev., 83, 488 (1951).
388. Barborik, Pracov. Lek., 19(1):11 (1967).
389. M. Barni, M. De Felice, and L. Reale, Polia Med., 2, 168 (1958).
390. F. Bartak, M. Tomecka, and O. Tomicek, Casop. Lek. Cesk., 81, 915-920 (1948).
391. E. Bary, Osp. Maggiore, 3:14, 18 (1926).
392. F. Berrod, Etude analytique et biologique de l'intoxication chronique par le baryum, centre de doc. univ. et S. E. D. E. S., Paris (1958).
393. P. L. Bidstrup and R. A. Case, Industr. Med., 13(4):260-264 (1956).
394. P. Bienvenu, C. Nofre, and A. Cier, Compt. Rend. Acad. Sci., 256(4):1043(1963).
395. T. W. Birch, L. I. Harris, and S. N. Ray, Biochem. J., 27:590 (1933).
396. H. P. Brinton, E. Frasier, and A. L. Koven, Publ. Health Rep., 67, 835 (1952).
397. E. Browning, Toxicity of Industrial Metals, London (1961).
398. Ethel Browning, Conference on Biological Aspects of Metal-Binding, 1960, University Park, Washington (1961).
399. R. Buchan, J. Industr. Hyg. and Toxicol., 30(1):10 (1943).
400. H. Buckup et al., Ztbl. Arbeitsmed., 6(1):1 (1956).
401. H. Buess, Helvet. Med. Acta, 17:104-136 (1950).
402. H. M. Carleton, J. Hygiene, 2 (1927).
403. H. H. Carney, Prac. Soc. Exper. Biol., 51(1):147 (1942).
404. Carozzi, Selenium, Hygiene du travail, 281, Geneve.
405. Carozzi, Tellure, Hygiene du travail, 9, Geneve.
406. N. Castelino, Ref. Arch. Industr. Health, 13(5):512 (1956).
407. J. Chack, Bymm, A. Chistoni, and E. Milamsi, Arch. Farm. Stern., 46, 147 (1921).
408. R. J. Chamberlin, Arch. Industr. Health, 19(2):231 (1950).
409. F. C. Christensen and E. C. Olson, Arch. Industr. Health, 16(1):8 (1957).
410. M. Chvapil and B. Cmuchalova, Nature, 186, 806 (1960).
411. J. H. Clark, Arch. Industr. Health, 20(2):117 (1959).
412. K. W. Cochran, I. Doull, M. Mazur, and K. P. Du Bois, Arch. Industr. Hyg., 1(6):637 (1950).
413. F. Czapek and T. Weil, Arch. exper. Pathal. Pharmak., 32, 438 (1893).
414. H. C. Cutter, W. W. Faller, I. B. Stocklen, and W. L. Wilson, J. Industr. Hyg., 31, 139 (1949).
415. T. A. Davies and H. E. Harding, Brit. J. Industr. Med., 7, 70 (1950).
416. R. H. De Meio, J. Industr. Hyg., 28, 229 (1946).
417. R. H. De Meio and W. W. Letter, J. Industr. Hyg., 30, 53 (1948).
418. R. H. De Meio, J. Industr. Hyg., 29, 393 (1948).
419. J. De Nardi, in: Pneumoconiosis, New York (1950), p. 82.
420. H. Desoille, C. Albachary, A. Rajdos, Rajdos-Todrok, Arch. Prof., 16, 185 (1955).
421. P. Dervillee et al., Arch. Mal. Prof., 23(9):598 (1962).
422. R. Doll, Brit. J. Industr. Med., 16, 181 (1959).
423. H. C. Dudley, Pub. Health. Rep., 53, 281 (1938).
424. C. C. Dundon and I. P. Hughes, Amer. J. Roentgenol., 63(6):797 (1950).
425. F. R. Dutra and E. J. Largent, Amer. J. Path., 26, 197 (1950).
426. F. R. Dutra, Amer. J. Path., 24, 1137 (1948).

427. W. Dutton, Vanadiumism, J. A. M. A., 56, 1648 (1911).
428. J. L. Eichgorn and P. Clark, Proceedings of the Seventh International Conference on Coordination Chemistry, Abstracts, Stockholm (1962), p. 126.
429. Egli, Marmet, Kapp, and Grandjean, Zs. Unfallmed. und Berufskrankh., 3, 210 (1957).
430. Engel, in: Work and Health of Arc Welders, Cited from S. V. Miller, Izd. Ukr. Tsentr. Inst. Gigieny Truda i Profzabolevanii, 24, 6-39 (1940).
431. J. Erbsloh, Arch. Toxikol., 18(3):156 (1960).
432. L. T. Fairhall, Industr. Toxicology, Baltimore (1949).
433. L. T. Fairhall, Brit. J. Industr. Med., 3(14):201-212 (1946).
434. L. T. Fairhall and R. Dunn, Publ. Health Bull., 293 (1945).
435. M. Fautrel, Arch. Mal. Profess., 19(1):5 (1958).
436. J. Ferin, Pracov. Lek., 8, 397 (1960).
437. J. Ferin, Ulrich, Menarik, Rabis, Pracov. Lek., 1, 35 (1960).
438. J. Ferin, Pracov. Lek., 1, 5-8 (1960).
439. O. L. Fitzhugh, A. A. Nelson, and C. I. Bliss, J. Pharm. Exp. Therap., 80, 289 (1944).
440. R. Flinn et al., J. Industr. Hyg. and Toxicol., 23(8):374 (1941).
441. H. Flintzer, Über gewerbliche Mannuergiftung, Diss., Jena (1930).
442. K. W. Franke and A. L. Moxon, J. Pharm. Exp. Therap., 61, 89 (1937).
443. K. W. Franke and A. L. Moxon, J. Pharm. Exp. Therap., 58, 454 (1936).
444. W. L. Fredrick and W. R. Bradley, Industr. Med., 15(8):482 (1946).
445. L. Friberg and A. Mystrom, Svenska lak. Ref. Arch. Industr. Hyg. and Occup. Med., 1, 539 (1953).
446. H. M. Garleton, J. Hyg., 2 (1927).
447. E. Ghislandi, Med. Lavoro, 48(10):566 (1957).
448. R. A. Gortner and H. B. Lewis, J. Pharm. Exp. Therap., 67, 358 (1939).
449. B. S. Gould, J. Biol. Chem., 232, 637 (1958).
450. I. G. Garcia, E. L. Garst, and W. E. Lowry, Arch. Industr. Health, 15(1):9 (1957).
451. J. Graca, F. Davison, and Feaveal, Arch. Environmental Health, 5(5):437 (1962).
452. H. B. Gray, J. Chem. Educ., 41, 2 (1964).
453. G. Guareschi and D. Boari, Alteneo Parmense, 19, 58 (1948).
454. J. E. Haine, Med. Off., 100, 113 (1958).
455. H. E. Hall, Brit. J. Industr. Med., 5(2):75 (1968).
456. A. Hamilton and H. Hardy, Industrial Toxicology, New York (1949).
457. J. Hamilton, Modern Phys., 20, 718 (1948).
458. R. Hansen, Ann. Chem. Pharm., 86, 208 (1953).
459. H. E. Harding, Brit. J. Industr. Med., 7, 76 (1950).
460. H. E. Harding, Brit. J. Industr. Med., 5, 75 (1948).
461. H. E. Harding and T. A. L. Davies, Brit. J. Industr. Med., 9, 73 (1952).
462. M. Harding, Lancet, 2(7043):393 (1958).
463. H. L. Hardy and I. R. Taber Shaw, J. Industr. Hyg. and Toxicol., 28, 197 (1946).
464. W. Haring, Dtsch. Med. Wschr., 67, 930 (1941).
465. H. E. Harrison, H. Bunting, N. K. Ordway, and W. S. Albrink, J. Industr. Hyg. Toxicol., 29, 302 (1947).

466. Q. Hartwig, T. Laffingwell, L. Melville, A. M. A. Arch. Industr. Health, 18(6): 505 (1958).
467. Q. L. Hartwig et al., Arch. Industr. Health, 18(6):505 (1958).
468. D. M. Hegsted et al., The Biologic, Hygienic and Medical Properties of Zinc and Zinc Compounds, Washington (1945).
469. Henke and Lubarsch, Handbuch Spez. Pathol. Anatomie und Histol., Vol. 3, Part 2 (1930).
470. L. Hermann, in: Lehrbuch der experimentellen Toxicologie, Berlin (1874), p. 131.
471. M. B. Hoagland, R. S. Grier, and M. B. Hood, Cancer Res., 10, 629 (1950).
472. W. C. Hueper, Arch. Industr. Hyg. Occup. Med., 6(2):187 (1962).
473. W. C. Hueper, Occupational Tumors and Allied Diseases, Baltimore (1942).
474. R. Jaksch, Med. Klinik, 13, 483 (1924).
475. Ickert, Staublunge und Staubtuberkulose (1928).
476. Kaulla, Münch. med. Wschr., 90: 399-402 (1943); Chem. Abst, 38(22):6399 (1944).
477. G. H. Keall, N. H. Martin, and K. E. Tunbridge, Brit. J. Industr. Med., 3, 175 (1946).
478. W. Kenneth and S. Cochran, Arch. Industr. Hyg., 6(6):637 (1950).
479. W. Kenneth, S. B. Cochran, J. Doull, et al., Arch. Industr. Hyg., 1, 637 (1950).
480. Kincaid, Strong, and Sanderman, Arch. Ind. Hyg. Occup. Med., 10(3):210 (1954).
481. E. I. King, C. V. Harrison, et al., Staub (Ref.), 18(10):327 (1958).
482. G. Klavis, H. Kohler, and F. Bister, Arch. Gewerbepath. Gewerbehyg., 14, 607 (1956); 15, 355 (1957).
483. F. Koelsch, Zentralblatt für Arbeitsmedizin und Arbeitsschutz, 9(2):33 (1959).
484. F. R. Koelsch, Handbuch der Berufskrankheiten, Jena (1959), p. 257.
485. F. R. Koelsch, Med. Klinik, 24 (1924).
486. Kotzing, Arch. Gewerbepath., 4, 500 (1933).
487. L. Kyker and A. Edgar, Arch. Industr. Health, 16(6):479 (1957).
488. G. C. Kyker and E. A. Grese, Arch. Industr. Health, 16(6):475 (1957).
489. R. E. Lane and A. Campbell, Brit. J. Industr. Med., 11, 118 (1954).
490. S. Laskin, R. Turner, and H. Stokinger, in: Pneumoconiosis, New York (1950), p. 360.
491. K. B. Lehmann and L. Herget, Chemiker Zeitung, 51 (1927).
492. K. B. Lehmann, Arch. fur Hygiene, 68(3):421 (1909).
493. L. Lenzi, Rass. Med. Appl. ol Lavoro Industr., 7, 301 (1936).
494. L. Lenzi, Ref. J. Industr. Hyg. Toxicol., 19(4):87 (1937).
495. C. E. Lewis, Arch. Industr. Health, 20, 455 (1959).
496. C. E. Lewis, Arch. Industr. Health, 19, 497 (1959).
497. C. L. Lindeken and L. Meadorso, Amer. Industr. Hyg. Assoc. J., 21(3):245 (1960).
498. L. Lindrichova, Pracov. Lek., 22, 1009 (1958).
499. A. Ch. Loken, Arch. Industr. Hyg. Occup. Med., 5(2):180 (1952).
500. K. D. Lundgren and H. Ohman, Virchow's Arch. Path. Anat., 325 (1954).
501. L. Machlin, P. B. Pearson, and C. A. Denton, Arch. Industr. Hyg., 6(5):441 (1952).
502. H. E. McMahon and H. G. Olken, Arch. Industr. Hyg., 1(2):195 (1950).

503. H. R. Marston, Physiol. Rev., 32, 66 (1952).
504. H. E. McMahon and H. G. Olken, Arch. Industr. Hyg., 1, 195 (1950).
505. L. McClinton and I. Schubert, J. Pharm. Exp. Therap., 94, 1 (1948).
506. I. C. McGowan, Arch. Industr. Health, 11, 315 (1955).
507. R. Meissner, Z. f. d. ges. exper. Med., 42, 275 (1924).
508. E. R. A. Meriwether, Some Problems of Industrial Hygiene and Occupational Disease [Translated from the English], Mir, Moscow (1960).
509. Ch. W. Miller, M. W. Davis, A. Goldman, and I. P. Wyatt, Arch. Industr. Hyg., 8:5, 453 (1953).
510. H. Minden and H. Thiele, Arch. Gewerbepath. Gewerbehyg., 16(4):396 (1958).
511. S. Moeschlin, Klinik und Therapie der Vergiftungen, Stuttgart (1969).
512. Molfino, Jr., Chemical Abstracts, 32(6):2244 (1938).
513. A. G. Morrow, Surgery, 28(6):1016 (1950).
514. Mott, Metal Carbonyls [Russian translation], Goskhimizdat, Moscow (1958). Cited from N. A. Belozerskii.
515. I. T. Mountain, E. R. Stockell, and H. E. Stikinger, Arch. Industr. Health, 12, 494 (1955).
516. I. Niekerk, Naunyk-Schmiedeberg's Arch. exp. Pathel. Pharmak., 184, 686 (1937).
517. C. Nofre, H. Dufon, and A. Cier, Compt. Rend. Acad. Sci., 257, 791 (1963).
518. Nuck, E. Remu, I. Holzmann, Hyg. und Infekt., 109:2, 598 (1929).
519. I. C. Paterson, J. Industr. Hyg. Toxicol., 29, 294 (1947).
520. L. Pauling, The Nature of the Chemical Bond, 3rd ed., Cornell University Press, Ithaca, N. Y. (1960).
521. L. Pauling, J. Chem. Educ., 39, 461 (1962).
522. R. Penalver, Industr. Med., 24, 1 (1955).
523. K. L. Peterson, Arch. Toxicol., 18, 160 (1960).
524. L. Prodan, J. Industr. Hyg., 14, 132 (1932).
525. A. Policard, Compt. Rend. Acad. Sci., 230, 899 (1950). Ref.: Arch. Industr. Hyg. Occup. Med., 5(1):90 (1952).
526. N. Otrzonsek, Int. Arch. Gewerbepath. Gewerbehyg., 24(1):66 (1967).
527. N. Otroznsek, Int. Arch. Gewerbepath. Gewerbehyg., 24(1):60 (1967).
528. D. Owen and H. Cohen, Lancet, 227, 989 (1934).
529. C. Reed, Excited States in Chemistry and Biology, Butterworths, London (1957).
530. C. E. Reed, Arch. Industr. Health, 13(16):578 (1956).
531. W. Reinl, Med. Klinik, 51, 1891 (1954).
532. W. Reinl, Arch. Gewerbepath. Gewerbehyg., 13, 721 (1955).
533. W. Reinl, Ref. Staub, 18, 143 (1958).
534. W. Reusert, Amer. J. Pharm., 56, 177 (1884).
535. F. N. Rhines, Phase Diagrams in Metallurgy (1960).
536. R. H. Rigdon, I. R. Couch, D. Brashear, and R. T. Qurech, Arch. Pathol., 59, 66 (1955).
537. B. Robertson and J. Hewitt, Biochim. Biophys. Acta, 49, 404 (1961).
538. J. Rodier, Arch. Mal. Profess., 16, 435 (1955).
539. J. Rodier, Arch. Mal. Profess., 2, 132 (1958).
540. J. Rodier, Maroc. Med., 37:395, 429 (1958).

541. P. Rose, Brit. Med. J., 1, 252 (1944).
542. S. N. Rosenthal, J. Pharm. Exp. Therap., 54, 34 (1935).
543. I. Rosenfeld and H. F. Eppson, Amer. J. Vet. Res., 18:68, 693 (1957).
544. H. Rothberg, L. A. Corallo, and W. H. Crosby, Blood, 14, 1180 (1959).
545. I. Sax, Dangerous Properties of Industrial Materials, Vol. 1, New York (1957).
546. P. Saccardo, Chimica, 11:411 (1955).
547. G. W. N. Schepers, Arch. Industr. Health, 12, 301 (1955).
548. G. W. Schepers, Arch. Industr. Health, 12, 134 (1955).
549. G. W. Schepers, Arch. Industr. Health, 12, 137 (1955).
550. P. Schuler et al., Med. Surg., 27(9):432 (1958).
551. L. Schwarts, S. M. Peck, K. E. Blacr, and K. A. Markuson, J. Allerg., 16, 51 (1945).
552. I. K. Scott, in: Pneumoconioisis, New York (1950), p. 262.
553. G. W. Shepers, Arch. Industr. Health, 12(2):140 (1955).
554. G. W. Shepers, Arch. Industr. Health, 12(2):134 (1955).
555. S. G. Sjoberg, Arch. Industr. Health, 11:505 (1955).
556. M. J. Smith, Proc. of the 6th Pareciffic Science Congress (1943), p. 167.
557. M. J. Smith, E. F. Stohlman, and R. D. Lillie, J. Pharm. Exp. Therap., 60, 449 (1937); Toxicol., 19(1):7 (1947).
558. H. Spannagel and Krefeld, Arbeitsmed., 28 (1953).
558a C. I. Spiegel et al., in: Pneumoconiosis, New York (1950), p. 326.
559. G. F. Sprague et al., in: Pneumoconiosis, New York (1950), p. 326.
560. Stadler, Ztschr. f. ges. Neurol. und Psych., 154, 62 (1935).
561. C. Steffee, Arch. Industr. Health, 20(5):414 (1959).
562 H. H. Steinberg, S. C. Massari, A. C. Miner, and K. Kink, J. Industr. Hyg., 24, 183 (1942).
563. J. H. Stermer and M. Eisenbud, Arch. Industr. Hyg., 123 (1951).
564. L. A. Stocken and R. H. S. Thompson, Physiol. Rev., 29, 168 (1949).
565. H. E. Stokinger et al., Arch. Industr. Hyg., 1, 379 (1950).
566. H. E. Stokinger and W. D. Wagner, Arch. Industr. Health, 17(4):273 (1958).
567. A. Stuart, Arch. exp. Pathol. Pharmakol., 18, 151 (1884).
568. F. W. Sunderman, Amer. J. Med. Sci., 236(1):26 (1958).
569. I. L. Svicbely, Biochem. J., 32, 467 (1938).
570. H. Symanski, Arch. Gewerbepath. Gewerbehyg., 9, 295 (1939).
571. L. Fairhall, The Toxicity of Molybdenum, Washington (1945).
572. L. Fairhall, T. Lawrence, and P. Neal, Industrial Manganese Poisoning, Washington (1943).
573. R. E. Tedeschi and F. W. Sunderman, Arch. Industr. Health, 16(6):486 (1957).
574. R. E. Tedeschi and F. W. Sunderman, Arch. Industr. Health., 16(6):480 (1957).
575. L. Teleky, Gewerbliche Vergiftungen, Berlin (1955).
576. D. Thomas and K. Stiebris, Med. J. Aust., 1, 607 (1956).
577. R. Truhaut, Les effets biologiques des thallium, Dermant, Paris 233 (1952).
578. L. Van Bogaert and M. Dallemagne, J. Industr. Hyg. Toxicol., 28(5):119 (1946).
579. G. Lavette, R. Cavier, and I. Savel, Arch. Internat. Pharmacodyn., 97, 241 (1954).
580. E. Vinke and H. Oelkens, Arch. exp. Pathol. Pharmakol., 188, 465 (1938).
581. H. C. Van Ordstrand et al., J. A. M. A., 129, 1084 (1945).

582. V. Ulehlova, Pracov. Lek., 5, 260 (1959).
583. L. Ulrich, Pracov. Lek., 4, 176 (1960).
584. D. W. Unseld, Klin. Wschr., 36(7):328 (1958).
585. R. I. P. Williams, Detoxication Mechanisms. The Metabolism and Detoxication of Drugs, Toxic Substances and Other Organic Compounds, London (1959).
586. C. H. Williams, Occup. Med., 4, 104 (1947).
587. R. H. Wilson, De Eds, and A. I. Cox, J. Pharm. Exp. Therap., 71, 222 (1941).